THE GENOMICS AGE

THE GENOMICS AGE

How DNA Technology Is Transforming
the Way We Live and Who We Are

GINA SMITH

AMACOM AMERICAN MANAGEMENT ASSOCIATION

NEW YORK * ATLANTA * BRUSSELS * CHICAGO * MEXICO CITY
SAN FRANCISCO * SHANGHAI * TOKYO * TORONTO * WASHINGTON, D.C.

Special discounts on bulk quantities of AMACOM books are available to corporations, professional associations, and other organizations. For details, contact Special Sales Department, AMACOM, a division of American Management Association, 1601 Broadway, New York, NY 10019.
Tel.: 212-903-8316. Fax: 212-903-8083.
Web site: www.amacombooks.org

This publication is designed to provide accurate and authoritative information in regard to the subject matter covered. It is sold with the understanding that the publisher is not engaged in rendering legal, accounting, or other professional service. If legal advice or other expert assistance is required, the services of a competent professional person should be sought.

Library of Congress Cataloging-in-Publication Data

Smith, Gina.
  The genomics age : how DNA technology is transforming the way we live and who we are / Gina Smith.
     p. cm.
Includes index.
ISBN 0-8144-0843-5
  1. Genetics—Popular works. 2. Genomics—Popular works. 3. Genetics—Social aspects.
  4. Genomics—Social aspects. I. Title.

QH437.S654 2005

                                                              2004012595

Printing number
10 9 8 7 6 5 4 3 2 1

CONTENTS

When most people hear the word eugenics, they think of the Nazis'
attempt in the 1930s and 1940s to murder their way to an Aryan
Germany. But few people know that eugenics—the pseudoscience of
genetically breeding humans—was first popularized decades earlier in
America. Eugenics was the first societal effort to manipulate genetics. Should we fear that a new eugenics is in the offing? What are the
ethical issues as the DNA sciences barrel into the future?

Every word you'll ever need to know to keep up with DNA researchers
and companies in the news, for investing and making societal and personal choices.

THE GENOMICS AGE

BEFORE WE BEGIN...

IT IS A GREATER achievement than the discovery of vaccines and antibiotics combined. And it is no exaggeration to say that, as a result of it, the world of human beings will never be the same.

I am talking, of course, about the discovery of the DNA double helix by an American and a Brit, James Watson and Francis Crick, in 1953. On a chilly February day, something profound happened. It barely got a mention in the papers that whole year. But Watson and Crick, they knew. "We found it!" Crick shouted upon bursting into The Eagle, an off-campus pub close to their University of Cambridge lab. "We have found the secret of life!"[1]

In April 2003, fully fifty years later, history was made again. A group of scientists announced they had taken Watson and Crick's great insight to yet another level. They published an enormous list—a list of the chemicals that make up all the genes in the DNA of the human race. In other words, they published the sequence of the human genome. And now the life-changing work can begin.

Knowing what a human being is made of is the first step toward knowing how to fix that human being when something goes wrong, or how to prevent something from going wrong in the first place. Eventually, it might even mean knowing how to build a better human being altogether. All of this is important, critical, even. But something also happened when this knowledge came to light. We humans—who are so happy with ourselves and our ability to reason, to investigate, to manipulate nature—became the first beings on the planet to take a look at ourselves at the most primary level, discovering the language in which our very existence is written.

The sum total of genes in a species—the DNA information that determines whether you have hair or hooves, teeth or a tail—is called the genome. Genomics is the emerging science of understanding the human genome, and of determining how the DNA in every human being affects identity, health, and disease. And genomics is launching other sciences almost as quickly as you can learn the terms. First functional genomics, then comparative genomics, then proteomics . . . the science breaks into subsets and into subsets again.

But one thing is certain. No matter how you slice and dice it, the new science of DNA will transform everything it touches: Medical treatment and diagnosis, especially. Criminology and genetic profiling. Cancer research and anti-aging. History. Ethics. Politics. And don't forget about the economy. Universities and businesses are sinking tens of billions of dollars into DNA-related fields.

"It's a giant resource that will change mankind, like the printing press," says James Watson, who should know.[2]

Johannes Gutenberg invented the movable-type printing press around 1450, and by the year 1500, there were a thousand books in Europe. That pace of change is generally considered to be extraordinary, but this DNA revolution puts that progress to shame.

In 1985, when I was an undergraduate studying chemistry at Florida State University, my organic chemistry professor told the class that the human genome wouldn't be mapped in our lifetimes. For a while, it looked like he was right. After all, the first genome—of the simple bacterium that causes meningitis—wasn't even decoded until 1995. It was tiny, and even that took years to do.

Then science turned a corner. Thanks mainly to advances in computer technology, researchers were able to outline the first draft of all three billion components of human DNA, about 200 New York City phone books worth of As, Cs, Ts, and Gs.

There still is an enormous amount of work to be done. Researchers are now trying to understand the contents of the book they have opened. According to Francis Collins, Human Genome Project leader, it is as if we have discovered the Book of Life, only to find the book is written in an unknown language. That means there is much left to do, and the benefits of the DNA sciences will arrive piecemeal, as we become increasingly fluent in its grammar and peculiar turns of speech.

And we must be careful not to get carried away with the hype surrounding this high-profile work. The tendency, says Collins, is to hear about the discovery of a new gene—such as a gene related to diabetes or heart disease—and immediately expect a cure for the ailment.

"Predictions in science tend to be over-optimistic in the short run," Collins told me as I was finishing up the first draft of this book. "But they tend to be under-optimistic in the long run. I think that applies here, too. Wildly overstated expectations of immediate benefits and [disease cures] from the Human Genome Project helped fuel the biotech frenzy of the late 1990s, but no one I knew thought that these expectations had any chance of happening at such a rapid pace.

"When the investment bubble burst," he added, "some people began to complain that the Human Genome Project was a failure and hadn't paid off. But it was the outrageous predictions that failed and didn't pay off. We will get there. It will happen. But not tomorrow or the next day. After all, it's one thing to derive the three billion letters of the code accurately and publicly. We've done that. But it will now require the best and brightest brains on the planet to go to the next level of understanding."[3]

But anyone wanting to put their excitement on hold because of that long to-do list need only look at extremely important genomics work that has already arrived. These results would've seemed like science fiction just a few years ago.

Consider. DNA evidence testing has proved the innocence of 144 convicted inmates—and counting—as of this writing.[4] It's cleared so many people on death row that, in 2003, then-governor George Ryan of Illinois commuted all the state's death sentences to prison terms of life or less.

Even historical crime mysteries are finding solutions. For instance, DNA evidence seems to have posthumously vindicated Sam Sheppard, who was accused of killing his wife in 1954. (You may remember the Sheppard case as the inspiration for the TV show and movie, *The Fugitive*.) The long-standing rumor that Thomas Jefferson fathered children with his slave, Sally Hemings, is now confirmed. Genetic tests show that some of the Hemings children were directly related to a Jefferson male.

And DNA evidence is being used to figure out everything from where Christopher Columbus is buried to whether Billy the Kid actually died in the1880s or, as rumored, lived on to be known as Brushy Bill, the elderly nursing home resident who, in the 1950s, claimed to be him.

The field of genetic testing is currently exploding, too. As researchers peg more and more gene mutations to specific disorders, DNA tests allowing you to be tested for them are right behind. You and your unborn child can already be tested for susceptibility for hundreds of diseases. In some case, finding out about a potential disorder and taking measures now to avoid it can save your life.

DNA medications are starting off more slowly, but they're coming, too. The startling effectiveness of DNA medicines such as Enbrel for rheumatoid arthritis and Gleevec for a certain kind of leukemia paints an optimistic future for medicines that precisely target the genetic problem behind a disease. And scientists believe they are on the threshold of creating personalized medicines—chemicals specially designed to work best with your particular genetic makeup.

The holy grail of the DNA sciences—the immediate tracing of every human disease and disorder to a single gene or group of genes—is further off. Yes, there has been progress in finding the genes linked to diseases such as cancer, heart disease, and diabetes. You'll read about a lot of that progress in this book. But it is certain to be more difficult than people once suspected. Most disorders aren't just mutations of a single gene, but many. And to treat genetic diseases, it will be necessary not only to understand the gene involved, but also the proteins the gene makes and everything that happens along the pathway from mutation to disorder. This will be the hard part.

Yet whether it takes years or decades, this much is certain: Medicine is forever changed. Because scientists now understand something about DNA, they are already using DNA knowledge to manufacture human hormones, help reduce heart blockages, shrink tumors, and treat multiple sclerosis. More developments are coming and, if history is any guide, they will greet us at a faster and faster rate.

Eventually, we will be living in a world where diseases are not just treated; they will be prevented from occurring in the first place.

Nobel Laureate David Baltimore told me that he had chills when he first read the paper that detailed the human genome. And he's seen a lot of biology in his long career. He is now the president of the California Institute of Technology.

Biology, he says, has entered a new era. "Instead of guessing about how we differ one from another, we will understand and be able to tailor our life experiences to our inheritance. We will also be able, to some extent, to control that inheritance. We are creating a world in which it will be imperative for each individual person to have sufficient scientific literacy to understand the new riches of knowledge, so that we can apply them wisely."[5]

Scientists such as Baltimore have long understood the frontier of the human genome and what it means to human beings. For the rest of us, it's taken a little longer. For most Americans, the science and terminology of the DNA revolution are brand new, just now appearing in the papers and on TV.

The science of DNA is simple, elegant, and ultimately graspable. You just need a little background in it, a little insight into who's doing what, what's coming, and what's just plain hype.

Cutting to the quick of the so-called DNA revolution is what this book is all about.

*   *   *

My goal with this book is to stick to developments likely to unfold in the next several years, detailing the advances that DNA research is expected to bring. That way, you can profit from the knowledge in your lifetime.

In the first three chapters, I'll get you familiar with the terms, techniques, and background you need to understand the rising tide of DNA stories in the news. If you don't know a gene from a chromosome—or if you just need a refresher on some newer terms and techniques—this section is for you.

Then, we'll take a look inside the labs, where key developments are happening in the hot areas of DNA fingerprinting, gene testing, cancer research, gene therapy, cloning and stem cell research, and anti-aging experimentation. In Chapters 4 through 9, you'll meet the minds behind the science, plus gain a plain English understanding of how they're taking on the challenge.

Finally, we'll reflect. Though I've included comments from ethicists and social scientists throughout, Chapter 10 digs deeper into the ethical issues facing us all. Should governments be permitted to compile DNA databases of each and every one of us? Could genetic testing result in an uninsurable and unemployable underclass? How will the DNA revolution affect your life and that of your family? I'll examine how current developments and their rush to reality will change the world for our children and our children's children.

These are issues we all need to think about. But without a decent grounding in the science of DNA—who the players are and what the technology is all about—the right decisions are difficult to make. You can't invest in or follow the DNA industry without knowing this stuff, either.

It's my hope that this book will give you not only the insight into what's happening in this historic revolution, but also the lay language and background to ask the hard questions—of yourself, the politicians who represent you, the business world, and the scientific community. There aren't too many other books that take on this challenge, but you've found one.

Now, onward!

IT'S WHO YOU ARE

RIGHT AROUND the time Washington crossed the Delaware River, the French chemist Antoine-Laurent Lavoisier wrote this in his notebook: "La vie est une fonction chimique."

Life is a chemical process.

Lavoisier was either lucky or prescient. (If he was lucky, it didn't last. French revolutionaries jailed and beheaded him in 1794.) But it was two centuries before scientists figured out the basic principles of heredity and came to widely accept that we inherit traits from our parents through a process that can only be called "chemical." Heredity is carried in our genes—genes that are made of DNA.

In the year 2000, scientists announced that they had launched what they said was a scientific revolution, that they had opened the book on human life. Three years later, in April 2003, they delivered the final version of that book.

They claimed they had figured out—chemical by chemical—what the DNA in human genes is made of.

"Essentially, we are now able to read our own instruction books," explains Francis Collins, director of the National Human Genome Research Institute in Bethesda, Maryland. And the term "instruction book," he says, hardly begins to define what the effort has uncovered. It is also a history book explaining how humans have evolved over time. It's a shop manual that describes with incredible precision how to build every cell in the human body. And most important, Collins says, it's a medical textbook containing insights that will help doctors predict and, eventually, cure disease.

> *It is humbling for me and awe inspiring to realize that we have caught the first glimpse of our own instruction book, previously known only to God.*
>
> *Dr. Francis Collins, Human Genome Project leader**

"We are the first generation in history to turn the pages of this book, an awesome and humbling experience for anyone to contemplate. In considering epic moments in human history, this has to be very high on the list. History will decide," he adds, "but I would place the Human Genome Project alongside splitting the atom or going to the moon."[1]

Introducing Your DNA

As most everyone knows by now, DNA is short for deoxyribonucleic acid. But do you know where your DNA is? Can you tell a gene from a chromosome? Did you know that your genes are located on your chromosomes and not the other way around? Do

---

* Collins's remarks, and others highlighted in this chapter, are taken from "What They Said: The Genome in Quotes," BBC News (June 26, 2000). This compendium of quotes from the public announcement of the completion of the rough draft of the human genome sequence is available at http://news.bbc.co.uk/1/hi/sci/tech/807126.stm.

you know how cloning might be used to fight diabetes, which companies are using worm DNA to figure out how to slow human aging, or how doctors are employing DNA knowledge to finally win the fight against cancer?

Most people don't.

The DNA sciences will dominate in the twenty-first century, and you need to understand the terms and the concepts if you're going to stay on top of and benefit from the huge DNA-related advances in medicine and other sciences.

At first, the science seems intimidating, but once you get a grasp of a few terms and concepts, you'll see it is all actually quite simple.

A View from the Top

You hear a lot about DNA "carrying" information—and I'll get to that in a minute. But first, let's talk about DNA as an object, an actual molecule that takes up physical space.

To give you some perspective, let's start big and get smaller.

Take a human body, any body. It consists, you may know, of ten systems: nervous, muscular, skeletal, endocrine, digestive, respiratory, circulatory, immune, reproductive, excretory.

> *I think we will view this period as a very historic time, a new starting point.*
> Craig Venter, founder of Celera Genomics

Each of those systems has organs. For instance, the stomach is an organ of the digestive system.

Every organ—like every living thing—is made up of cells. The stomach is made of stomach cells.

Almost every cell, stomach or otherwise, has a nucleus at its center. And this is, for me, where things get interesting.

Every nucleus includes chromosomes, rod-like structures that, under a microscope, most resemble bundles of thread. Every cell's nucleus contains exactly twenty-three pairs of these chromosomes. (The exception to this is reproductive cells, which contain half the normal amount of chromosomes. That makes sense considering that reproduction is the result of the fusing of two cells—a sperm and an egg.)

Look closer at any particular chromosome—let's choose chromosome 19 inside the particular nucleus of a stomach cell we're examining—and you'll find that chromosome's DNA tightly coiled up inside. If you unraveled that DNA and straightened it, you would find that it is shaped very much like a ladder. Sugars and phosphates form the sides of each ladder, and the four so-called "bases" pair up to form the rungs.

> *This is the outstanding achievement not only of our lifetime, but in terms of human history. I say this because the Human Genome Project does have the potential to impact the life of every person on this planet.*
>
> *Dr. Michael Dexter, director of The Wellcome Trust*

The bases are guanine, adenosine, thymine, and cytosine—G, A, T, and C for short. You may also hear them called *letters* or *nucleotides*. If you think of DNA as a language, and I do, then this is the alphabet.

A given gene (made of DNA) is simply a given group of base pairs on a DNA molecule. For instance, here on chromosome 19 you can find a long string of bases that together form the so-called APOE gene. If you were unlucky, you may have inherited a dangerous variety of this gene (there are three varieties) on your chromosome 19. This could affect your ability to break down cholesterol and fat—leading to coronary heart disease, Alzheimer's, or other fat-related ailments.

However, and this is where the gene sequence could come in handy, if you were able to discover this risk factor early, through a genetic test, you might choose to go easy on the bacon double cheeseburgers, a choice that could extend your life.

> ### WHY YOU AREN'T A BLUE BLOOD
>
> For centuries, presumably all the way back to Aristotle, folk-lore had it that heredity passed through our blood. Think of the terms "bad blood," "mixed blood," "royal blood," "blue blood," or "bloodline" and you get the idea.
>
> The irony is that there is no heredity coded in your red blood whatsoever. The red blood cells are the only kind of cells in your body that don't have DNA—because they're the only cells in your body that don't have nuclei.
>
> Go figure.

To summarize, you have about 30,000 genes located throughout your twenty-three pairs of chromosomes, which are found in the nucleus of almost every one of your cells. These genes describe, in the alphabet of Gs, As, Ts, and Cs, everything about you—from how tall you are, to how curly your hair is, to how likely you are to suffer from bad breath or cancer.

Your personal DNA sequence is the language in which everything about you is written. Interestingly, almost every cell in your body has all the information required to build an entirely new you.

Visualizing Your DNA

A single DNA molecule is incredibly long and skinny. Uncoiled from a microscopic chromosome, a single strand would stretch about two inches. String out all the DNA from all twenty-three chromosomes from a human egg just about as big as the comma at the end of this clause, and its length would add up to about six feet.

Line up all the threads of DNA from every cell in your body end to end, and the entire length would be long enough to reach to the sun and back 500 times. But the same strand would be many thousands of times skinnier than a human hair.

> *In the deepest sense, DNA's structure and function have become as much a part of our cultural heritage as Shakespeare, the sweep of history, or any of the things we expect an educated person to know.*
>
> Microbiologist Ross L. Coppel, from his book with G.J.V. Nossal, Reshaping Life: Key Issues in Genetic Engineering (Melbourne, Australia: Melbourne University Publishing, 2002)

Now consider how compact your DNA is. Almost every cell in your body contains more than six feet worth of DNA coiled up inside.

Even so, the standout feature of DNA is the way it stores information—information that precisely instructs the cell how to replicate itself and what functions to perform.

But What Does DNA Do, Exactly?

DNA's job is simple. Its code tells your body how to build protein.

Protein is at the foundation of all living things. All living cells depend upon proteins for virtually all their products and processes. Cells—whether they're bacteria, plant, or animal cells—use proteins for a variety of processes, from fighting infection, to sending and receiving messages, to rebuilding damaged parts.

You, as a human being, contain at least 50,000 different kinds of protein. And each kind of protein has a specific job to do. There are structural proteins for building your hair (collagen), your skin (keratin), and your ligaments (elastin). Hormonal proteins like insulin carry messages and regulate body processes. The hemoglobin in your blood is one example of a transport protein. There are antibody proteins to protect your cells against invaders and protein enzymes for digesting and otherwise breaking things down. The list goes on and on.

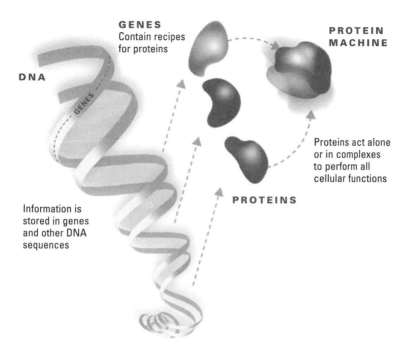

FIGURE 1-1. How DNA codes for protein.

The primary job of DNA is to tell the body what proteins to build and how to build them. The order of the chemical bases A, T, C, and G on a gene gives the cell the recipe for that particular protein. Scientists used to think that one gene always directed the body to create one protein, but now they know that a single gene can potentially create more than one kind of protein.

> *We share 51 percent of our genes with yeast and 98 percent with chimpanzees—it is not genetics that makes us human.*
> *Ethicist Dr. Tom Shakespeare, University of Newcastle*

The idea of DNA creating proteins is a critical one. Proteins are the workhorses of the human body; they do all the work in a cell. They carry out chemical reactions, form new tissue, send signals between bodily systems, regulate body chemistry, you name it. The simplest way to think of DNA is that it is, at its most basic, just a huge, long file in which all the instructions for creating the proteins in your body are written down.

> *We have to focus on the possibilities, develop them, and then*
> *face up to the hard ethical and moral questions that are*
> *inevitably posed by such an extraordinary scientific discovery.*
>
> United Kingdom Prime Minister Tony Blair

Eric Lander of the Massachusetts Institute of Technology has called this the fundamental secret of life. "The secret of life is this huge diversity of components; fifty thousand [proteins] that are all specified in the same simple description of the DNA language."[2]

When you hear that scientists "have mapped" the human genome, this is what they are talking about. They have figured out the exact order in which A, T, C, and G appear on human genes. Quite obviously, there are areas of variation that explain why one human has blue eyes, for instance, and another brown. But the DNA of any two humans is well over 99 percent identical.

DON'T IT MAKE MY BROWN EYES BLUE (OR NOT)

People have been wondering for thousands of years why it was that their baby had hazel eyes, when the parents had blue and brown.

A gene, recall, is a given stretch of DNA located on one of your twenty-three chromosomes. That gene codes for a specific protein, which in turn performs a specific function or helps build a certain structure.

The gene for brown eyes, for instance, codes for a protein (an enzyme, actually) that selectively deposits pigment on the irises of your eyes. If you have blue eyes, you lack that protein.

Now scientists are in the process of figuring out which proteins are coded for by various sequences of bases (genomics), what those proteins do (a field called proteomics), and what happens when the sequence goes awry (functional genomics). There is also a separate effort going on to map the variations between the DNA in people, so we can discover the exact sequences that account for our differences. One of the ultimate goals is for scientists to develop medicines that precisely address mutations in the code—supplying missing proteins when necessary and eventually modifying the code altogether.

> *We've now got to the point in human history where for the first time we are going to hold in our hands the set of instructions to make a human being.*
>
> *Sir John Sulston, Nobel Laureate*

How DNA Stores Information

Probably the easiest comparison to draw is between the way computers store information and the way DNA does it.

Computers deal in 1s and 0s. Everything they do is translated down to that level. For instance, if you save my name "gina" in your word processor, your computer would translate those letters into a stream of 1s and 0s. "Gina" would translate into:

01100111011010010110111001100001

You can think of DNA, on the other hand, as using a four-letter alphabet. As an example, a short sequence of bases on a gene might look something like this:

AAATTGCGCCCAATACGTACGTTTACGA

Recall that this sequence of letters—representing the four bases—A, C, T, and G—is a recipe. It tells the cell exactly what proteins to make. One error—that is, one single change, or mutation, in the sequence—and the gene might not make its protein, or make the wrong one altogether. Sometimes that doesn't matter. Other times, it's critical.

WHAT MAKES YOU UNIQUE?

If you compared your DNA sequence to mine, you'd have a hard time finding differences between us. You could go literally thousands of letters before finding a single difference—say, a T instead of a C.

It turns out that all of us are incredibly alike. The As, Gs, Ts, and Cs in your DNA appear in anyone else's DNA in the same order about 99.9 percent of the time.

But is that enough difference to account for the vast individuality of the human species?

It is. Recall that we each have twenty-three pairs of chromosomes—one set arrived intact from your mother, the other from your father. And your parents inherited those from their parents.

Take your chromosome pair 1—your largest pair of chromosomes. The one from your mother is either the one she received from her mother or the one she received from her father. Let's call those 1m and 1f. From your father, you also receive a chromosome that came from each of his parents; let's call these 1M and 1F.

So at birth, your pair of chromosome 1 could be 1m1M, 1m1F, 1f1M, or 1f1F.

And this same recombination happens with every chromosome you have.

All of this may explain why you have your father's nose and your maternal grandmother's hair, while your sister has just the opposite. From this example, it might seem as if you only have genetic material from one grandparent on each side, but that is overly simplistic. Remember, your father's DNA is a combination of his parents, and that your mother's is a combination of hers.

That is why, unless you have an identical twin, there is no one on the planet exactly like you, and there never will be.

There are thousands of hereditary diseases that result from just a single letter error in a sequence. Perhaps, because of a typo during copying, there now is a G where a T should be, or there is a repeat of a sequence of Cs over and over.

But because DNA information is inherently linear and digital in nature, reading it and, eventually, manipulating it becomes merely a matter of refining technique.

"This kind of digital information is the easiest kind of information to manipulate, which is why, to my mind, genomics is the central science of biology," Caltech's David Baltimore told me. "You can do so much with it."[3]

It's a giant resource that will change mankind, like the printing press.

Dr. James Watson, Nobel Laureate and co-discoverer of the double helix

As you will see in this book, the digital information that is your genome can be used to identify someone at the scene of the crime, or to identify whether you're at risk medically and need to take precautions against hundreds of known diseases. Doctors will use the code to create personalized, side-effect-free medicine that works better for you than anyone else, or to help infertile couples identify which of their in vitro fertilized eggs is healthiest. Down the road, scientists hope to use DNA knowledge to lessen the effects of aging and, eventually, lengthen lives by several years or more.

Years from now, stem cell technologies, though controversial, hold great promise for treating dozens of serious diseases, as do methods of gene therapy, a field involving the actual tinkering of genes to help cure certain diseases.

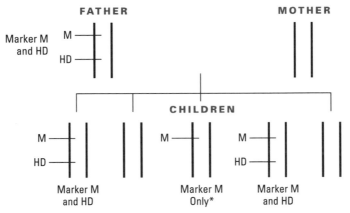

* Recombinant: Frequency of this event reflects the distance between genes for the marker M and HD.

FIGURE 1-2. How genes are passed on from parents to child.

All of this is only possible because scientists are now beginning to understand the stuff we're made of. Progress is coming in fits and starts—and there are many frustrations along the way.

Francis Collins, the leader of the Human Genome Project, was the scientist who discovered the cystic fibrosis gene in 1989. Still, there's no cure. But he's optimistic.

I never thought it would be done as quickly as this.

Fred Sanger, Nobel Laureate and inventor of DNA sequencing

"Finding the gene, hooking it up with a particular disease, gives you immediate insight into what the actual molecular problem is," Collins told an audience on CNN. "It gives you almost immediately the ability to predict who's at risk for that disease, and in some instances that in itself can be life-saving. If you know, for instance, you're at a high risk for colon cancer, well, you go in and you get screened for that disease and you pick that up while it's still easily treatable, and that's a home run."

Collins added: "That's terrific, that's what we hope for. But not all diseases allow you that kind of intervention. There are many steps that you have to follow then before you can harvest, from this wonderful information about the gene, how to put that [information] into practice in the medical arena. But you can't do that harvest if the gene information is not available to you."[4]

We now have the possibility of achieving all we ever hoped for from medicine.

Lord Salisbury, United Kingdom science minister

David Baltimore elaborates further: "Just knowing the genetic defect will help us better understand how we treat [a disease]. And if we can't treat it—and it may be a long time before we can treat [the diseases associated with many mutations], we'll be able to say, Hey, there's no mystery here. Here are the lifestyle changes you ought to think about to reduce your chances of getting this."[5]

Finding the specific mutations that make people more likely to get a disease will give doctors a clue about what might happen to you before it's too late.

"The idea is that we can use the blood as a window to look into the body and distinguish between health and disease," says Leroy Hood, inventor of the automatic sequencing system most labs use to find genetic defects.

ONE GENETIC ERROR TOO MANY?

Single gene errors account for more than 4,000 known hereditable diseases, and the list is growing rapidly. Your risk for such diseases as cystic fibrosis, sickle cell anemia, Lou Gehrig's disease, and Huntington's disease now can be determined by looking at the DNA from any of your cells through a microscope.

For instance, in Huntington's disease, the triplet CAG on chromosome 4 is repeated too many times—CAG occurring more than forty times in a row seems to result in the disease. The result is a flawed protein that ends up interfering with the function of nerve cells.

Remember, though, that all our chromosomes are paired, so we have two copies of each gene. Some genetic diseases— let's use cystic fibrosis as an example—are recessive. That means they do not develop unless a person has two flawed copies of a gene. The normal one simply acts as a backup.

Other diseases, like Huntington's, are dominant. That is, getting just one mutated copy of the gene from your mother or father is enough to predispose you to a certain disorder.

How serious a mutation is and whether it is enough to cause a disease, of course, varies. Diseases such as cancer involve many mutations across several genes and even several chromosomes.

Now that the final sequence of the human genome is available, the challenge for researchers at universities and private companies is using that information to find the multiplicity of genetic problems behind cancer, heart disease, diabetes, and other major killers. The next challenge will be finding treatments relevant to a range of mutations and gene products. No one should underestimate the size of the job that lies ahead.

In the near term, genomic sciences and other sciences that look at what exactly is happening in a cell, or an organ going awry, will change everything. "It is going to move us from worrying about being sick to worrying about remaining well," Hood told me. "That alone will increase the average lifespan of a person by ten or fifteen years."[6]

Hundreds of companies nationwide are rushing to find correlations between genetic mutations and specific disease. For quite obvious reasons, just finding the typos in DNA has become a multibillion-dollar industry.

It's All in the Matching

I've already said that the actual shape of the DNA molecule is like a ladder—a double-twisted ladder. Recall that the two sides of the ladder are long chains of sugar and phosphate. The rungs are pairs of chemicals—*base pairs*, we call them.

There is a rule for how bases pair up. Cs always pair with Gs, and Ts always pair with As. Always. No exceptions. In other words, if one strand says ATCGATCG then, because of the matching rule, the other strand automatically says TAGCTAGC.

> *It would surprise me enormously if in twenty years the treatment of cancer had not been transformed.*
>
> *Dr. Mike Stratton, Cancer Genome Project leader*

When scientists James Watson and Francis Crick discovered this matching rule in 1953, there was a lot of excitement—because finally there was a theory describing how it is that cells divide into other cells that look and act just like them. (The theory also explains why a man and woman have a human baby—as opposed to, say, a kitten.)

The matching scheme allows cells to replicate into exact copies of themselves. When the cell divides, the DNA molecule unravels into its two individual strands.

One cell gets ATCGATCG. The other gets TAGCTAGC.

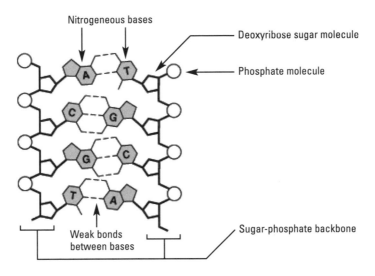

FIGURE 1-3. Bases and how they pair.

Both unpaired DNA strands now mix in the chemical soup inside the cell, attracting their complementary base pairs. So you now have two daughter cells where once there was one. And each daughter cell ends up with a copy of DNA with its two strands:

ATCGATCG
TAGCTAGC

The helical shape of DNA explains how the incredibly long string of chemicals fit in the cell, how DNA divides and puts itself together again, and how it is that such a repetitive string of the same four chemical letters could possible specify the code for all the amazing diversity in life and the human body.

It represents an immense step forward for humanity in deciphering the makeup of life itself.

Yoshiro Mori, former Japanese prime minister

Watson and Crick won the Nobel Prize for their work in 1961.

Thanks to the matching rule, there is a DNA industry. Because of it, Fred Sanger and, later, Leroy Hood were able to develop their sequencers, and splicing and recombining DNA became possible.

Understanding how bases always match up allowed high-powered computers to shred and put back together human DNA and figure out exactly what the human genome sequence is.

Moving On

There is so much about DNA that I couldn't include here, information that could fill (and does fill) entire college-level biology textbooks.

I didn't mention, for instance, that a chemical called RNA (short for ribonucleic acid) is responsible for reading the codes found in DNA and bringing the code to the cellular organs required to actually build the designated proteins. I didn't mention that many genes don't code for proteins at all, but instead are stopping and starting signals the RNA needs to know when a gene on a given stretch of DNA begins and ends. And there is lots of "junk" DNA that is of unknown importance.

> *The deciphering of the Book of Life is a milestone in science.*
> Roger-Gerard Schwartzenberg, former French research minister

I didn't talk about ribosomes or mitochondria or any of the myriad of cell mechanisms involved in genetic processes.

I'll work some of that into coming chapters. But for now, consider the bases covered.

It's a Fact

Fact: A simple list of the bases of all the DNA in your genes— the As, Cs, Ts, and Gs—would fill about 200 New York City phone books. That's about three billion letters.

Fact: Most people think of meat when they hear the word *protein*. And rightly so. You're made of it—about 50,000 to 100,000 different kinds of proteins comprise the human body and perform all its functions. In other words, protein really is meat—the meat that is you!

Fact: It has been called the "central dogma" of molecular biology: DNA makes RNA makes protein. That is, the sequence of bases in DNA tells the sequence of bases in RNA how to put together a complex three-dimensional protein molecule. As dogmas go, this isn't a good one—there are exceptions, it turns out—but "DNA makes RNA makes protein" is an old saw that most beginning scientists find helpful.

Fact: The term *chromosomes* means "colored bodies." That's because thready chromosomes easily absorb the dye scientists pour on cells before examining them under a microscope. Scientists identified chromosomes long before they had any idea what critical role they played in DNA replication.

Fact: It's true that every cell in your body has the same three billion or so genes; but obviously, all cells are not alike. How is this? Each cell has many more genes than it uses. Some are turned on (expressed) and others are turned off (unexpressed). Figuring out why and how cells express some genes but not others is a central question—still unanswered—of the DNA sciences.

CHAPTER TWO

HOW WE GOT HERE

IN 1995, scientist Craig Venter published the first genome of an organism, the genome of a bacterium that causes a rare form of meningitis. That genome included about 1,743 genes—or 1,830,137 base pairs. Scientists think of this today as an utterly tiny genome, a DNA sliver, really. But it was a huge feat at the time. The effort floored the scientific community.

Within five years, Venter was CEO of the high-flying private company, Celera Genomics, and locked in a race with the government-sponsored Human Genome Project to sequence all 30,000 genes and 3.2 billion base pairs that make up the entire human genome. The two sides declared a draw in 2000, when they jointly unveiled the first draft of the human genome, leaving scientists all over the world with the task of making sense of the data. (The final draft—considered the finished release—was announced in April 2003, coinciding with the fifty-year anniversary of Watson and Crick's discovery of the DNA double helix.)

Despite breathless expectation and gales of media hype around millennium time, the mapping of the human genome turned out to be not so much an answer to a question as a new question that's led to countless others.

Scientists say those new questions mark the beginning of an era.

"This is the genomic era, and everything from now on is fundamentally different from what came before. Biology can really start now," says David Galas, chancellor and chief scientific officer of the Keck Graduate Institute in Claremont, California. Charming and articulate, he takes quick issue with those who say that the genome race was little more than hyped-up big science. "We may not understand it all yet, but we now have everything we need to know to understand the scope of the problem. People say it was overhyped, but I don't think it can be overhyped. How can you overhype the fact that biology is just beginning?"[1]

> *The theory is confirmed that the pea hybrids form egg and pollen cells which, in their constitution, represent in equal numbers all constant forms which result from the combination of the characters united in fertilization.*
>
> Gregor Johann Mendel, the father of genetics, 1866

Galas told me that he compares the blueprint of the human genome to Dmitri Mendeleyev's 1869 presentation of the periodic table, which finally showed the relationship of elements to one another.

"There was chemistry before the periodic table. They had discovered oxygen, for instance," Galas adds. "But they didn't understand how the bonds between chemicals worked. They couldn't come up with plastics or silicon alloys or any of the materials made possible by an understanding of the periodic relationship between chemicals."[2]

So in ten years, we went from knowing virtually nothing about the genes that make a human being to understanding almost everything. Now, scientists are left with the task of figuring out exactly what all these genes do and the type and shape of proteins they make.

It's a huge leap in the space of a decade, and we'll spend the rest of this book explaining how the human genome sequence will directly affect you. But, as the old saying goes, you can't understand where you're going until you understand where you've been. This chapter explains how we got here.

The Fruitful Search for Sperm and Egg

You could say—and I will say—that practical genetics began when civilization first started domesticating animals. Herders started to selectively breed animals to choose traits they wanted to see in their animals, but they didn't really understand how their breeding experiments worked.

Almost all the ancient Greek philosophers took a shot at figuring it out. Aristotle, predictably, thought all heredity came from the father—the mother was responsible only for providing the less intelligent raw material. Pythagoras thought pretty much along the same lines.

Empedocles, accounting for why children sometimes resemble their mothers, thought Pythagoras wrong, saying that male semen and female fluids blended to create offspring. But the theory fell apart when he explained why sometimes a child looks like neither parent. It might, he wrote, have something to do with the things mothers looked at during pregnancies, such as sculptures and statues.

And then there was the idea of spontaneous generation, which most everyone hung on to during the Middle Ages and even afterward. The belief was that living beings could arise from nonliving matter. Flies were born from rotten meat, they thought. Insects were born of stagnant ponds.

The scientist who did the most to destroy the prevailing theory of spontaneous generation was Antonie van Leeuwenhoek, a Dutch amateur scientist who was the first to use a magnifying lens to see sperm cells in 1679. (Judging from drawings in his lab notebook, he also spied bacteria and viruses.) Unfortunately, he wrongly concluded that sperm cells were exclusively responsible for reproduction, and that each little sperm encased a fully formed model of an organism, but in miniature. He called them "little animalcules."

Another scientist, William Harvey, concluded just the opposite. Harvey, who was the pampered on-call physician of King Charles I, had lots of spare time and energy to think about it. After studying chickens, he determined that all life comes from the egg: "Ex ovo omnia," he proclaimed. He believed, correctly, that even mammals had eggs, and he led a search for them that failed spectacularly. It wasn't until the early 1800s that scientists finally located the mammalian egg for which Harvey had looked so hard.

In 1875, a scientist named Oscar Hertwig made the discovery that set science straight. A single sperm has to fertilize a single egg before reproduction can occur. In his famous statement describing the merging of the two nuclei inside the female egg, Hertwig wrote: "Es entsteht so vollständig das Bild einer Sonne im Ei." *(It arises to completion like a sun within the egg.)*[3]

Genetics rose like a sun right along with it.

A Monk Named Mendel

The son of German farmers, Gregor Mendel performed so brilliantly in school that his parents felt they had no choice but to give him the best education they could afford: life as an Augustinian monk.

The man now known as the father of modern genetics was a nature lover who spent hours tending the monastery garden. And it was there that, working with just the peas in his garden, Mendel guessed, with incredible accuracy, some of the facts later proven by Watson, Crick, and others many decades later.

An avid student of botany and agriculture, Mendel in 1856 selected twenty-two varieties of culinary peas (he called them his "children") to crossbreed in his garden. He claimed he did it "for the fun of the thing." And over the course of eight years, he had fun studying more than 10,000 plants—meticulously following traits such as seed color, plant height, and pea color.

By observing these successive generations of pea plants growing in his garden, Mendel came up with a set of theories that turned out to be dead-on and are now known as the Mendelian laws of inheritance.

WHO DISCOVERED DNA?

It wasn't Gregor Mendel, the nineteenth-century monk who grew peas in his garden to figure out heredity. It wasn't James Watson and Francis Crick, either. They won the Nobel Prize for determining the double helix shape of DNA, a critical observation. But they didn't discover DNA.

The discoverer of DNA was a young Swiss chemist named Friedrich Miescher. In 1869, he was examining the excretions on discarded medical bandages, the biological phenomenon commonly known as pus. Miescher called the milky substance that he detected "nuclein." Almost a century later, scientists figured out its importance and renamed it DNA.

The high point of his theories was that plants and animals pass along what Mendel called "discrete factors" to their offspring. Also, he claimed that all plants and animals inherit half of these factors from their mother and half from their father. Factors, he said, never blended. Then, using a theory of what traits were dominant over the others, he even figured out statistically how characteristics would pass from generation to generation.

We now call Mendel's "factors" genes. And, in fact, they come half from the mother and half from the father; they move intact from generation to generation; and they never blend—just as the uncelebrated monk predicted.

Mendel set out all these theories in a paper in 1865, but it generated little interest. He was absolutely unrecognized in his lifetime. According to his biographers, Mendel was confident he was on to something and was fond of repeating, "Meine Zeit wird schon kommen." *(My time will come.)*[4]

It took thirty-five years, but Mendel's time truly did come. And it's still here. His short discourse, *Experiments in Plant Hybridization*, is widely considered to be among the most important scientific publications of all time.

On the Origin of Species

At about the same time that Mendel was planting peas in his garden, Charles Darwin was sailing the world as a naturalist on board the British naval ship *HMS Beagle*. He was gathering material for his book, *The Origin of Species*, which, on its first day of sale in 1859, would sell out of all copies printed.

> *Man is developed from an ovule, about 125th of an inch in diameter, which differs in no respect from the ovules of other animals . . . . The embryo itself at a very early period can hardly be distinguished from other members of the vertebrate kingdom.*
>
> Evolutionist Charles Darwin, 1871

As most everyone knows, Darwin's book (and its successor, *The Descent of Man*) made the point that all life on earth is the result of natural selection, that it evolved from the simplest one-celled forms, and that all life-forms are somehow linked. Before his publication, common reason had it that man was distinct from the other animals and had not changed since the dawn of time.

After Darwin, educated people the world over began to widely accept that we also are animals shaped by nature's forces over long spans of time. The discovery of the DNA molecular structure as the vehicle of earthly heredity further rammed that theory home.

ODE TO BACTERIA

Unless you're a scientist, you probably think of bacteria as a nasty thing you want to avoid. Yet these tiny microbes are among the most valuable tools available to researchers, ranking up there even with the most high powered of computers.

A bacteria culture is useful to anyone hoping to create recombinant DNA (i.e., a string of DNA that scientists paste together from different sources). For instance, scientists create artificial human insulin by inserting the correct sequence of bases coding for DNA into a bacterium's DNA sequence. Then they set the bacteria into a medium, and let the bacteria do their job of producing insulin.

Bacteria make this especially easy because, unlike most life-forms, their DNA isn't concealed inside their nucleus. Rather, bacterial DNA is free floating, whirling around on independent little wheels (called plasmids) inside their one-celled bodies. Also, bacteria can easily transfer their DNA to each other, either by pushing it out and sucking it up in the medium they're swimming in, or through viruses that carry foreign DNA from one bacteria species to another. That makes it easy for researchers to transfer genes into one species of bacteria and to easily spread it around to many others.

If that doesn't make you respect the little guys, consider this: Bacteria have existed on the planet longer than anything else. In fact, they were probably the earth's only tenants for close to two billion years. That's one reason why bacteria are so diverse. As Eric Grace points out in his book, *Biotechnology Unzipped*, you are more similar to a potato or a shark than one species of bacteria is to another.

Scientists not only appreciate that diversity, but they like the flexibility. Bacteria multiply quickly, they're cheap, and they're not picky about where they live. Not surprisingly, scientists use them to create proteins such as insulin, not to mention vaccines, hormones, and many other products.

Watson, Meet Crick

On February 28, 1953, Francis Crick and James Watson burst into an off-campus pub and announced to the lunchtime crowd that they had found the secret of life. Well, as the saying goes, it ain't bragging if it's true. The secret they were talking about—the secret they won a Nobel Prize for unveiling—was the structure of DNA. And it was in fact a secret of life, if not *the* secret.

The two researchers, working together in a lab for several years, had managed to figure out the double helical shape of DNA and how its two strands lined up with one another. Understanding the shape and replication method of DNA paved the way for the biotech and genomic revolutions that came later, and helped explain countless questions thinkers had posed before.

But the key question for him, says James Watson, was in a 1944 book by theoretical physicist Erwin Schrödinger, titled *What is Life?* In the slim volume, Schrödinger said something Watson says astounded him. Schrödinger postulated that we could understand heredity if we better understood atoms, and that life was described in a kind of "genetic code" (he coined that phrase) that lies in the exact configuration of our molecules. But where was the code, and how did it work?

"It polarized me," says Watson. "It changed me from wanting to be a naturalist like Charles Darwin [to] wanting to be a geneticist searching for the secret of life."[5]

Watson and Crick weren't the only ones looking. In the decades before they began their work at Cambridge University, momentum was building in the United States and abroad in the search to discover how heredity worked.

> When the full map of the human genome is known . . . we shall have passed through a phase of human civilization as significant as, if not more significant than, that which distinguished the age of Galileo from that of Copernicus, or that of Einstein from that of Newton. . . . We have crossed a boundary of unprecedented importance . . . . There is no going back. . . . We are walking hopefully into the scientific foothills of a gigantic mountain range.
>
> Ian Lloyd, House of Commons member, 1990

In 1869, young Swiss chemist Friedrich Miescher had discovered DNA in the pus in surgical bandages and named it "nuclein," or nucleic acid. Though he didn't understand its importance or structure, he was first to pinpoint the substance. He called it nucleic acid because it was an acidic substance he found only in the nucleus of cells.

Three-quarters of a century later, a British scientist named Fred Griffith figured out the significance of Miescher's discovery. In a 1928 experiment, Griffith discovered that if you mixed two kinds of bacteria (one pneumonia-causing, one not), one strain would cause the other strain to take on infectious properties. Griffith concluded that there was a "transformative principle" at work that, in essence, was genetic material. Still, no one knew what that genetic material was.

Finally in 1944—the same year Schrödinger released *What Is Life?*—Oswald Avery answered the question in his New York City lab. He'd spent years grinding up bacteria and eliminating possibility after possibility in a search of Griffith's transforming principle. Finally, he ended up with the nucleic acid, DNA. It was, he claimed, the genetic material carrying the information of heredity.

The race to figure out exactly how DNA worked was officially on.

The Race Was On

"My interest in DNA had grown out of a desire, first picked up while a senior in college, to learn what a gene was," Watson says with characteristic pluck. "It was certainly better to imagine myself becoming famous than maturing into a stifled academic who had never risked a thought."[6]

At Cambridge, the American Watson and his lab partner, Francis Crick, were working at Cavendish Laboratories. Neither was officially supposed to be working on the problem of DNA, but both had an avid interest in it. Particularly, they were fascinated by the work of physicists Maurice Wilkins and Rosalind Franklin, who were trying to use X-ray technology to come up with a possible explanation for how DNA was arranged. So Watson and Crick worked their day jobs and pursued their obsession with DNA secretly, following Wilkins and Franklin's work closely.

> *Nature prevails enormously over nurture when the differences of nurture do not exceed what is commonly to be found among persons of the same rank of society and in the same country.*
> *Francis Galton, anthropologist and eugenicist, 1876*

At the same time, the famous biochemist Linus Pauling was said to be close to discovering DNA's structure at the California Institute of Technology. (His son Peter, a graduate student at Cavendish, regularly tormented Watson and Crick with news of his father's progress.) It looked as if Pauling, well financed by Caltech and not limited to after-hours study, was going to get there first. But then something happened.

One summer night in 1952, Watson developed a negative of an X-ray he'd snapped of some DNA. "The moment I held the still-wet negative against the light box," says Watson, "I knew we had it. The telltale helical markings were unmistakable."[7]

A year later, though, Watson and Crick were no further along than that. They knew that DNA was helical—shaped like a twisted spiral—but they didn't understand how many strands the helix had or how its components (the bases A, C, T, and G) fit together and replicated. Then, Peter Pauling walked into their lab with bad news: His father had come up with the structure of DNA. But when Peter showed his father's manuscript to them, they were intensely relieved: Pauling had outlined a three-strand helix, an arrangement Watson and Crick had already ruled out as impossible.

"Basically, Pauling had fallen flat on his face," says Watson. "We knew then that we had our chance."[8]

On the Home Stretch

Watson and Crick knew immediately that it was only a matter of time before Pauling realized his error. So they appealed to their boss at Cavendish, William Bragg, to let them openly study DNA and get to the answer first.

Bragg gave them a temporary reprieve, and Watson and Crick started a feverish run. First, they visited the lab of Wilkins and Franklin at King's College to get a look at what their latest X-rays of DNA looked like. To their amazement, they found X-rays, already a year old, that not only confirmed their suspicion of a helical structure but also suggested that the number of strands could be set at two.

Immediately, they began building models to illustrate what they found. Still left to figure out was exactly how the hydrogen, sugars, phosphates and bases were aligned. A friend convinced them to pair C with G and A with T, given that those chemicals always appeared in equal proportions in DNA (the so-called Chargaff ratio). Crick and Watson suddenly realized exactly how to build their seven-foot-tall wood and wire model—with the sugar-phosphate arms of the ladder along the sides and the base pairs making up the rungs.

"The final attack... only took a few weeks," says Crick. "Hardly more than a month or so after that our paper appeared in *Nature*."[9]

It is difficult to resist the fascinating assumption that the gene is constant because it represents an organic chemical entity. This is the simplest assumption that one can make at present, and since this view is consistent with all that is known about the stability of the gene it seems, at least, a good working hypothesis.

Nobel Laureate scientist Thomas Hunt Morgan, 1926

The photo of the gawky young scientists standing beside their eight-foot-tall model of the DNA double helix is one of the most famous of the twentieth century. The wire and peg model is twisted into the shape of a double-spiraled ladder.

"The double helix instantly proposed the solutions to two of the problems posed by Schrödinger," says Watson. "How do you store and carry genetic information?"[10]

"Instantly, we knew that DNA genetic information must be carried by the order of the four bases—A, G, T, and C—along with the sugar-phosphate background. So just the order of the bases gave the information. DNA information was encoded in a sort of digital-like way. In turn, copying involved separating the double helix's two strands. The resulting single strands then serve as molds, as templates for the formation of second strands using the base pair rules. Opposite an A, you have to have a T; and opposite a G, you have to have a C."[11]

FIGURE 2-1. Watson and Crick walking along the Backs, 1953.
(Courtesy of the Cold Spring Harbor Laboratory Archives.)

In their first article about their discovery in the scientific journal *Nature*, Watson and Crick made what many consider to be the understatement of the century: "It has not escaped our notice that the specific pairing we have postulated immediately suggests a possible copying mechanism for the genetic material."[12]

The double helix did, however, escape the notice of newspapers at the time. Journalists missed the story. Hardly anyone covered it. That says something intriguing about science journalism, for sure. Today, most scientists consider the Watson-Crick discovery to be the most important discovery of the century, and possibly even of the millennium.

Decades later, Crick wrote in his memoirs that the ideas behind the double helix were "ridiculously easy, since they do not violate common sense."

"I believe," he continued, "there is a good reason for the simplicity of the nucleic acids. They probably go back to the origin of life . . . At that time mechanisms had to be fairly simple or life could not have started. The double helix is indeed a remarkable molecule. Modern man is perhaps 50,000 years old . . . but DNA . . . [has] . . . been

around for at least several billion years . . . yet we are the first creatures on earth to become aware of its existence."[13]

> *Even at birth the whole individual is destined to die, and per-*
> *haps his organic disposition may already contain the indica-*
> *tion of what he is to die from.*
>
> Sigmund Freud, 1924

Watson and Crick's discovery won them a Nobel Prize in 1961,
along with Maurice Wilkins. Rosalind Franklin, unfortunately, died
before the prize was awarded.

The Post Watson-Crick Era

"What is truly revolutionary about molecular biology in the post-
Watson-Crick era is that it has become digital," writes Richard
Dawkins in his book, *River Out of Eden.* "The machine code of the
genes is uncannily computer-like."[14]

By digital, Dawkins means that our genes are comprised of a digital code that directly translates into something else. In the case of
genes, we're talking about amino acids. These are the building blocks
of protein.

THE GENETIC CODE CHEAT SHEET

Scientists figured out, in 1967, how DNA specifies the
building of a protein.

Recall that in every life-form, the letters A, C, T, and G
(i.e., the bases) perform the same function. They build proteins by instructing another chemical, called RNA, to put
the proteins together one building block after another.
RNA substitutes the T (thymine) for U (uracil) in all cases.
This is an oversimplification, but you get the idea.

The building blocks are called amino acids, and there are
precisely twenty of them. Three-letter "words" of DNA are

called codons, and they are the same in any living cell they appear in. Here's a cheat sheet of amino acids and their corresponding three-word codons.

Amino Acid	Codons
Phenylalanine	TTT, TTC
Leucine	TTA, TTG, CTT, CTC, CTA, CTG
Serine	TCT, TCC, TCA, TCG, AGT, AGC
Proline	CCT, CCC, CCA, CCG
Isoleucine	ATC, ATA, ATT
Methionine	ATG
Threonine	ACT, ACC, ACA, ACG
Valine	GTT, GTC, GTA, GTG
Alanine	GCT, GCC, GCA, GCG
Cysteine	TGT, TGC
Tryptophan	TGG
Tyrosine	TAT, TAC
Arginine	CGT, CGC, CGA, CGG, AGA, AGG
Histidine	CAT, CAC
Glutamine	CAA, CAG
Asparagine	AAT, AAC
Lysine	AAA, AAG
Glycine	GGT, GGC, GGA, GGG
Aspartic acid	GAT, GAC

? Glutamic acid   GAA, GAG
?
?
? End code (code that signals the end of an instruction) TAA,
?
? TAG, TGA

Even the Swiss biochemist Miescher, the scientist who discovered DNA in the nineteenth century, imagined that perhaps a string of chemicals could carry a hereditary message, in the same way that roughly twenty-five to thirty letters of the alphabet can explain all the concepts in all the world's alphabetic languages.

It turned out that all twenty amino acids—the twenty amino acids required to build the some 50,000 proteins found in the human body—are coded for by just the four bases A, C, T, and G. After Watson and Crick, the main question was how.

Two young scientists, Har Gobind Khorana and Marshall Nirenberg, came up with the answer in 1967. They figured out that if you take the four bases and put them together in groups of three, you can put them together in sixty-four different arrangements ($4 \times 4 \times 4 = 64$). That's more than enough to code for all twenty amino acids.

The question of interest is no longer whether human social behavior is genetically determined; it is to what extent.

Pulitzer Prize–winning author and biologist Edward O. Wilson, 1978

The code is simple. Three letters form a "word"—a so-called codon—that indicates a specific amino acid. For instance, the codon CAG tells the cell to assemble the amino acid glutamine. Some proteins are very small, requiring only a couple of hundred amino acid building blocks, while others require thousands. Insulin, for instance, is an example of a tiny protein. It is a chain of just fifty-one amino acids. It is no accident that insulin was among the first man-made human proteins—it was the easiest to assemble.

Scientists altered the genome of a bacterium to include the exact combinations of As, Cs, Ts, and Gs that human pancreas cells use to create insulin. Once they were able to do this, they could just let the successive generations of bacteria do their work. Millions of diabetics today use genetically engineered insulin, and that's a direct result of the genomics revolution.

Discovering how to assemble man-made insulin was a major revolution in the world of diabetics. Before that, pig insulin was used, and not all diabetics could tolerate it.

What a Difference Three Letters Make

In the 1970s and 1980s, scientists began piecing together, base by base, how various hereditary diseases lined up with what they were learning about genetics.

One of the first diseases looked at was Huntington's chorea, a dominant human genetic disease that is particularly horrible. By dominant, I mean that, unlike many genetic diseases, you only have to inherit one copy of a mutation to get it. That is, you don't need it from both your parents, only one.

In 1978, the musician Woody Guthrie died of this neurological disorder, an inevitably deadly disease that slowly robs its victim of neurological function over the course of fifteen to twenty-five years. Guthrie's widow joined up with a doctor named Milton Wexler, a man who knew the ravages of the disease intimately. His wife had it, his three brothers-in-law had it, and his daughters, Alice and Nancy, each had a 50 percent chance of getting it.

Wexler became obsessed with finding the gene responsible. "I became an activist because I was terribly selfish," he says. "I was scared to death that one of my daughters would get it, too."[15]

Wexler's daughter, Nancy, took up the fight, too. Everyone told her to forget it. Finding one gene in potentially a hundred thousand (they did not know how many human genes there were at this time) was a crazy goal; it just couldn't be done.

But Nancy Wexler persisted. Following a tip about a large extended family in Venezuela that was suffering from Huntington's, she flew there and started interviewing people. She discovered a woman who had been affected by the disease who had 9,000 descendants. Of these descendants, 371 had the disease and more than 1,500 shared the same 50 percent risk of having at least one affected parent.[16] Then, she began collecting blood.

These were "hot, noisy days of drawing blood," Nancy Wexler later wrote.[17] Progress was slow. In 1983, a doctor Wexler was working with had managed to "locate" the Huntington's problem on the short arm of chromosome 4. He thought it was somewhere in a region of about a million bases in length. Eight years after that, there was no further progress. "The task has been arduous in the extreme, in this inhospitable terrain at the top of chromosome 4. It has been like crawling up Everest over the past eight years."[18]

Within ten years, before a [newborn] child leaves the hospital the parents will have the option of having the genome profile on CD-ROM.

Craig Venter, genome sequencing pioneer, 2003

Then, in 1993, researchers finally found it—a gene on chromosome 4 gone badly awry. The repetition of the three-base word CAG (spelling glutamine) was the cause of the problem. And the more repeats, the worse off an affected person is. The cursed number turns out to be about thirty-nine. Thirty-nine CAGs in a row and you end up with your first Huntington's symptoms at age 66. If you have fifty repetitions of the word, you'll begin to lose brain function right around age 27.

Matt Ridley, in his book *Genome*, puts it this memorable way: "If your chromosomes were long enough to stretch around the equator, the difference between health and insanity would be less than one extra inch."[19]

The story about the Huntington's gene is important to understand for several reasons. For one thing, it illustrates how, in the space of just a few years, medical science went from knowing practically nothing about the syndrome to understanding its causes in minute detail.

The main passion that led to the double helix was curiosity.
And Crick and I focused our curiosity on DNA.

James D. Watson, Nobel Laureate and
co-discoverer of the double helix, 2003

It also drives home a sobering realization. Though we can test anyone for the presence of the Huntington's mutation, there is, to this date, no cure for it, no way to delve into the millions of brain cells carrying this mutation and fix them.

Nancy Wexler and her sister have so far declined to be tested. Both of them are a decade older than their mother was when she was diagnosed. "When we were trying to develop a test, we assumed we'd both take it. Then once the test existed, we were thinking about it differently. Our family talked an enormous amount about the consequences. Even if you live at risk all of your life, and you've thought about it and cried about it, there's a certain amount of denial that helps you get through the day. Being tested can take that away."[20]

Automatic Sequencing

What finally made the identification of the Huntington gene and countless others possible—in fact, what made the mapping of the entire human genome possible—was the invention in the 1970s of an automated way to sequence DNA.

Before automated methods, sequencing DNA was an overwhelming task. The tiny DNA molecule is easily damaged, and reading one microscopic A, C, T, or G out of millions (or, in the case of the human genome, billions of letters) made many people believe that the task wouldn't be completed in our lifetimes. According to statistics published online by the Human Genome Project, it would take one scientist, typing sixty words a minute, eight hours a day for fifty years to type out a human genome, base by base.

It wasn't until the mid-1970s that more than a few scientists were able to determine the sequence of any strand of DNA longer than eighty bases. Then came the so-called Sanger method for DNA sequencing, named after its inventor Fred Sanger. It involved, first, the shredding of a DNA sample into many different-size pieces and making many copies of each one so that scientists would have a wealth of DNA strands to test. Then, the Sanger method attached a radioactive tag to the last A, C, T, or G in each segment. A researcher would go in and "read" those tags later. The method immediately accelerated sequencing from the rate of about 150 base pairs per researcher per year to about 1,500 per year. But the method was costly, messy, inefficient, and labor-intensive. Sanger won the Nobel Prize in chemistry for this work in 1980.

Do you want to know when you are going to die, especially if you have no power to change the outcome?

Nancy Wexler, biologist and gene hunter, 1992

Caltech researcher Leroy Hood raised the bar yet again in 1986 with his invention of a partially automated sequencer, one that attached fluorescent dyes to the bases. Outputs were much easier to read—a machine could accomplish it—which sped up the task even more. This method was also safer because it didn't use the potentially hazardous radioactive chemicals.

Like the original Sanger method, Hood's automatic sequencer worked by tearing apart DNA at every possible point. For instance, take the strand:

ACATGCGTAGTCAGTAC

It would be torn into pieces of various size and then arranged according to length, as follows:

A . . . AC . . . ACA . . . ACAT . . . ACATG . . . ACATGC . . . ACATGCG . . .

But Hood came up with what turned out to be a better idea. He imagined what would happen if you assigned a color to each of the bases. If each fragment took on the color of the base in the last position, an inexpensive four-color computer printer could deliver the results on paper, eliminating the need to have a technician reread the sample after the chemicals marked them.

For example: If A were blue, C were green, T were yellow, and G were red, then each of those pieces in the previous example would have a color assigned to it. (Just the terminating base of a sequence piece has to be color-coded.) So, in the previous example, that would be:

A	C	A	T	G	C	G
Blue	Green	Blue	Yellow	Red	Green	Red

By printing out the color associated with the last base in every possible length sequence, you get the exact sequence of bases. It is a simple but extraordinarily effective way of speeding up DNA sequencing.[21]

There would be no genomics without the ability to store, compare, analyze, search, and annotate all of the sequences generated in the genomic age.
Harold Varmus, Nobel Laureate and cancer research pioneer, 2002

In 1985, using current methods of DNA sequencing, it was possible to sequence only about 25,000 base pairs per person per day. In 2004, it is possible to sequence more than 12,000 base pairs per

FIGURE 2-2. Caltech biologist Leroy Hood, pioneer in automated DNA sequencing.

second. And thanks to improved computer technology, that number is growing all the time.

"This is one of the most exciting times in biology," says Hood. "The revolutions that have been generated by the [Human Genome Project] have barely been felt, but there is one profound change that has already occurred, and that is the realization that biology is fundamentally an information science."[22]

To summarize, once researchers came up with an automated way to discover the sequence of bases in a stretch of DNA, mapping the human genome finally became a reality. There is no way that manual sequencing could have yielded such quick and accurate results. The method of tearing a piece of DNA into pieces allows a computer to put it together in the same way you would put together a jigsaw puzzle—corners first, for instance, then the sides, and then adding each piece one at a time if it fits the overlapping pattern around it. The speed and single-minded precision of a computer is ideal for this task. Technology, then, paved the way for Celera and the Human Genome Project's race to map the human genome. By the end of the project, computers were decoding the human genome to the tune of 12,000 base letters a second.

Craig Venter Against the World

The mapping of the human genome, as Craig Venter, founder of Celera Genomics, has been quick to point out, was as much a triumph of computer technology as it was of biology.

"As a biologist, I didn't know anything about high-end computing. But fortunately, I am an experimental scientist, because I had to evaluate all the major computer manufacturers in the world to try and work out which computer might be able to assemble the human genome," says Venter.

"There was no way to sort out the claims from IBM, Digital, Sun, and Silicon Graphics, so I gave them a problem to solve. I gave each of them the Haemophilus genome [sequenced earlier] and our algorithm and asked them to see if they could improve on the eleven days it took us to assemble it with a Sun [Microsystems] 32-bit computer. Only two computers, IBM and [Compaq] Digital, could even run the experiment. With some optimization with the Alpha chip, Digital got it down to nine hours. Eleven days to nine hours was a big improvement. So we worked with Compaq to build a massive facility."

> *We set up a large sequencing factory. It took about six months to build our facility and totally equip it. Our laboratory is the size of football fields . . . we have substituted electrons for people and initiated very high throughput efforts.*
> *Genome sequencing pioneer Craig Venter, 2002*

Celera ended up with 1,200 Alpha processors, the fastest computer technology available at the time. Reporters who visited the Celera lab in the late 1990s always commented on the eerie environment, unlike any lab they'd ever seen: just a few researchers and two football-field-lengths' worth of large computers pumping the data out. The method that Celera relied on was "whole genome shotgun sequencing." Essentially, the Celera researchers took all the DNA out of the cells (from three anonymous males and two females) to be tested and cut it into various-size fragments. They made copies (cloned) those DNA fragments and then let the computers go to work by mathematically reassembling the pieces and naming each base.

"Imagine trying to line up forty-five million of those sequences in terms of working out where the overlaps are, especially when there are a lot of repeats in the genome. So we only put things together when there was a single mathematical solution in the entire human genome. There was less than one chance in ten to the fifteenth [10^{15}] of making an error. That is why we were so confident that this would work even when everybody else was saying it was absolutely impossible," says Venter.[23]

And, in fact, most other scientists around the world regarded the Celera effort with amusement. Sequencing a bacterial genome quickly with the shotgun method was one thing, but completing the human genome? After all, more than a thousand researchers at the U.S. government's Human Genome Project had been hard at work on the same problem since 1988.

> *Now, for the first time, we have an historical anthology of ourselves, some of it passed down for a billion years. We're just learning how to read the story, and it's sure to enthrall us for decades to come.*
> *MIT scientist Eric Lander, 2001*

In fact, the heated rivalry between Venter and the Human Genome Project (originally headed by James Watson) began in 1991. Venter, then a scientist for the National Institutes of Health (NIH), was using a method called "expressed sequence tags" to quickly find genes and label them, rather than the meticulous and lengthy experiments scientists normally used. Venter told a congressional committee that the NIH was patenting his discovered genes at the rate of about 1,000 a month, a comment that enraged Watson. Watson complained that Venter's approach of merely locating genes without determining function was "cream skimming" that "virtually any monkey could do."[24]

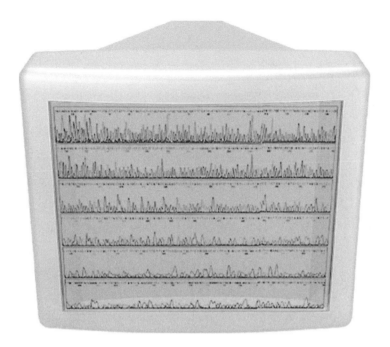

FIGURE 2-3. Computer with DNA sequence info on screen.

Watson ended up in a war with his boss at the NIH, Bernadine Healy, and lost his job over it. Francis Collins, a scientist and gene hunter who played a lead role in discussing the cystic fibrosis gene years earlier, was appointed to take his place.

When Venter left the NIH in 1991, a venture capital firm invited him to try out his gene sequencing strategy at a nonprofit, The Institute for Genomic Research (TIGR). And in 1998, Venter dropped yet another bombshell. He was teaming up with the giant PerkinElmer, Inc. to create a new company using a high-powered automatic sequencing machine. The company would pump out the sequence in just three years, he promised, and for only $300 million.

For the next three years, it was battle by press release, with each side, Celera and the federally funded Human Genome Project, continually announcing that it had sequenced this or that percentage of the genome. Finally, the two sides reconciled in early 2000, calling a truce to the race and simultaneously announcing at a June 2000 White House event the rough draft of 3.2 billion letters of the human genome. Together, they delivered the results of both efforts to a scientific journal.

Moving On

What's next? Everything.

Despite all the enthusiasm, there remains an enormous amount to do. Mapping the genome is just the beginning.

Scientist Leroy Hood gave me a handy comparison. He says knowing the sequences of the human genome is like owning a Spanish dictionary. "You might have all the words, but you still need to speak the language [to use it]."[25]

Caltech's David Baltimore couched it this way: "The sequencing of the genome is an icon of science. But it is a largely symbolic. There is a lot more to do."[26]

Knowing the chain of chemicals that makes up human DNA only gets you so far. Scientists have now turned their attention to figuring out what comes next—actually interpreting the code.

> *Its significance would be comparable to that of the effort that led to the conquest of space, and it should be carried out with the same spirit. Even more appealing would be to make it an International undertaking, because the sequence of the human DNA is the reality of our species, and everything that happens in the world depends on those sequences.*
>
> Nobel Laureate Renato Dulbecco, imagining the Human Genome Project in 1986

That's the knowledge scientists need for building drugs that act on particular genes and proteins, not to mention rearranging components on the genome itself. How long it will take researchers to turn the human genome sequence into a quantifiable revolution in health care is still debatable. So far, success has been spotty. Then again, progress comes in fits and starts.

Mark Walport, a director at The Wellcome Trust, has pointed out that researchers discovered a gene called BRAF in June 2002. That gene is reportedly involved in more than 70 percent of malignant melanomas. By April 2003, companies were already working on drugs to target it.[27]

It's a Fact

Fact: In 1985, when the Human Genome Project was first proposed, many critics thought it was absurd. At the time, the technology did not even exist to decode the sequence of a simple bacterium, much less a human being. The largest genome yet sequenced was the tiny Epstein-Barr virus.

Fact: Sequencing the human genome and its three billion base pairs seemed sheer boredom to many scientists. Sydney Brenner of the U.K.'s Medical Research Council jokingly recommended that governments hand felons the job: The worse the crime, the bigger the chromosome they would have to sequence.

Fact: You've heard of genomics, the study of DNA. But have you heard about all the other fields it's spun off? There's proteomics, the study of the proteins that DNA codes for and what they do. There's structural genomics, a field that aims to generate 3D images of proteins so that drug companies can better target them. And there's comparative genomics—the analysis of human DNA alongside of the DNA of monkeys, mice, fruit flies, bacteria, and other common model organisms—which is a key strategy for figuring what individual genes do.

Fact: In 1996, Pope John Paul II suggested that "an ontological discontinuity" lies between apes and humans, the point where God inserted the human soul. Given that humans have one less chromosome than apes—scientists say two ape chromosomes fused to become our chromosome 2—author Matt Ridley wryly suggests a location for the soul. "Perhaps," he says, "the genes for the soul lie near the middle of chromosome 2."[28]

Fact: In 1989, President George H. W. Bush awarded the National Medal of Science to recombinant DNA pioneers Stanley Cohen and Herbert Boyer, he referred to the Human Genome Project as the Human *Gnome* Initiative.[29]

Fact: While studying a small virus, Fred Sanger and colleagues reported that genes don't necessarily have to lie one after another; they can actually overlap. Physicist Freeman Dyson likens this insight to a Mozart duet where two violinists stand facing each other with a single piece of music between them. One violinist plays normally, while the other plays all the notes from the bottom of the page up. "I like to call Fred Sanger's virus the Mozart virus," says Dyson. "It shows that nature can compose a genome as cleverly as Mozart could compose a duet."[30]

Fact: Scientist Max Delbruck once joked that Aristotle should've been awarded a posthumous Nobel Prize for anticipating DNA. The Greek philosopher had argued that the "form" of a chicken lies in an egg, and that an acorn held within it the plan of an oak tree.[31]

Fact: Gene sequencing heavyweight Craig Venter named his company Celera Genomics from the Latin *celeris,* meaning rapid or swift. The company's slogan: "Speed matters. Discovery can't wait."

CHAPTER THREE

YOUR GENOME—
AN OWNER'S MANUAL

YOU HAVE twenty-three pairs of chromosomes that make you uniquely you. But among the 3.2 billion base pairs on those chromosomes, you have only about 31,000 genes.

That alone is one of the biggest surprises of the Human Genome Project. We have about a third fewer genes than anyone expected—and not even double the amount of genes a roundworm has. There is a variety of amoeba that has nearly 200 times the amount of genes that we do.

It's easy to understand why this bothers people. It's humbling. But the question of why our genome is smaller than other "simpler" animals is just the tip of it. Yielding puzzles, insights, and new evidence supporting or tearing down what we thought we already knew, the human genome sequence doesn't disappoint. This short chapter details some of the goodies.

We Are More Alike Than We Thought

The difference between you and that unrelated guy walking his poodle down the street is just 0.1 percent. That means you share with any given unrelated person 99.9 percent of the same DNA. And, in fact, the order of all those bases is exactly the same in almost all people.

As it turns out, only a small fraction of the genome—about three million bases—is different from person to person. Of course, you are yet more similar to your relatives.

To picture this, imagine that everyone's individual genome is a book. You and another person have almost identical books—same story line, same order of chapters and words. But on some random page—say, page 100—your book has a typo. Maybe it says "their" when it should say "there." And perhaps you have a word repeated—two "the's"—whereas the other person has only one "the."

According to Human Genome Project statistics, if you and a friend started reciting your DNA sequences at the rate of one letter a second, it would be more than eight and a half minutes before the two of you reached a difference.

> *We all are essentially identical twins.*
>
> Craig Venter, human genome sequence pioneer

And scientists say that just a few thousand of those differences are responsible for the biological differences between you and me. Craig Venter, founder of Celera Genomics, the company that led the private effort to map the genome, took that notion to the extreme when he told a BBC interviewer: "That means we all are essentially identical twins—even more than we thought."[1]

This point may seem counterintuitive. After all, the variation between individuals seems vast—there are blondes and brunettes, the tall and the short, the blind and the seeing. Some people live to be a hundred, while others die in childhood of dreaded heritable diseases.

A person with just a few base letters difference will likely get Huntington's chorea, while another will not. Another person, with just a single gene change, will die of cystic fibrosis at an early age. Another will inherit a single mutation that predisposes her to an aggressive form of breast cancer.

But the fact is, of the three billion base letters making up the human genetic code, only three million of them (0.1 percent) are unique for each person. That turns out to be enough to account for differences in looks, vulnerability to disease, and countless other traits, but it is nevertheless a much smaller variation than anyone expected.

We Are More Like Chimpanzees (and Yeast) Than We Thought

If you are more like your poodle-walking neighbor than you thought, consider the chimpanzee.

You have roughly 99.1 percent of your genes in common with the chimpanzee, our closest relative on earth. That means, if you analyzed the DNA of a human and a chimp side by side, you would discover that most of the material is indistinguishable.

The overlap between a mouse and a human is surprisingly close, too. We have nearly 75 percent of our genes in common. With roundworms, we have about a 40 percent overlap.

And about a third of the genes in yeast also show up in human DNA.

Aristotle had an idea he called *scala naturae*, also known as "the ladder of life." He supposed that all life was related, and that it was possible to place it all on a continuous scale, from lowest to highest life-form. While this idea has led to a lot of misconceptions—for instance, the Nazis used the "some lives are worth more than others" idea to promote the murder of millions of Europeans—the concept of a continuous chain of beings on this planet turns out to be absolutely true.

Commonality of genes across every life-form in the planet is a key discovery, scientists agree. It provides final evidence that humans weren't created and didn't evolve separately from everything else. Rather, all forms of life on the planet are intricately related, and by studying DNA, it is possible to determine exactly when and how certain life-forms spun off.

> With the genomic revolution, new tools have become available to study human diversity at the DNA level. With these tools we have been able to reconstruct human history with a surprising degree of clarity.
>
> *Scientist Douglas Wallace*

"We evolved through this effort of billions and billions of years, working back from single-cell organisms to more and more complex organisms," Craig Venter told CNN. "We have the same genes as in the bacteria. The enzymes that correct defects, the genetic code from radiation damage, UV damage in a bacteria, are the same ones that are related to cancer in humans. Those processes are highly conserved.

"So, in fact, the best hope for understanding human biology and medicine is that we can use the genomes, the sequences from other species, to understand the human ones."[2]

There Is No Such Thing as Race

And speaking of common ancestors, there turns out to be no such thing as race at the DNA level.

"Race as used in the United States is a social and political construct derived from our nation's history," wrote Celera researcher Harold Freeman. "It has no basis in science. The biological concept of race is now believed to be untenable."[3]

In other words, you cannot tell simply by looking at someone's DNA whether they are black or white. Genotype (the description of a person's DNA) should not be confused with phenotype (what they actually look like). "From a genetic perspective, all humans are Africans, either residing in Africa or in recent exile," adds anthropologist Svante Paabo of the Max Planck Institute of Evolutionary Anthropology in Leipzig, Germany.[4]

Scientists tell us, in fact, that DNA among humans is more similar than the DNA of many other kinds of animals, mostly because we are such a young species, from an evolutionary standpoint. As a species, it turns out we're extremely closely related to each other. We've discovered, for instance, that the difference between any two random chimpanzees is about four times greater than the difference between any two randomly selected humans.

The tenth of a percentage of DNA that is different from human to human even existed when we were all black Africans, about 100,000 years ago, and there were only about 10,000 humans on the planet. The variation between humans back then hasn't increased at all.

> *Race as used in the United States is a social and political construct derived from our nation's history. It has no basis in science. The biological concept of race is now believed to be untenable.*
>
> Biologist Harold Freeman

"There are now over six billion human beings on the planet, distributed from the Arctic Circle to Tierra del Fuego," writes scientist Douglas Wallace in a far-reaching essay about DNA and human history. "They exhibit striking differences in physical features, indicating adaptation to different environments."[5] The genomic revolution, he says, gives us powerful new tools to reconstruct human history. Wallace says that, using mitochondrial DNA (a type of DNA carrying only information from the maternal line) and paternal Y chromosomes, it is now possible to see exactly how and when various populations migrated out of Africa.

A Lot of the DNA in Our Cells Is "Junk"

Junk is an unkind and increasingly inappropriate word for it, but the truth is, scientists don't know exactly what the long stretches of repetitive DNA (usually long stretches of Gs and Cs) in our cells are for. And so for now they call it "junk DNA." But the fact remains that about 95 percent of the DNA in our chromosomes doesn't include genes at all.

Some of the junk DNA is undoubtedly the remnants of viruses that embedded their DNA into ours, says Caltech's David Baltimore. He explained to me that many viruses work by transcribing—actually, the term would be *reverse transcribing*—their DNA into the DNA of the life-forms they infect. That is why, he says, that some human DNA looks like "a sea of reverse-transcribed DNA," with just a few regular genes thrown in.[6]

Junk DNA likely also has a lot of regulatory material in it, adds George Stamatoyannopoulos, the young founder of Regulome Corp. in Seattle. He claims that simply identifying the 30,000 genes in the genome is not enough. What's needed is a close look at the rest of the material.

From a genetic perspective, all humans are Africans, either residing in Africa or in recent exile.

Anthropologist Svante Paabo

"People have done a lot of work in looking into genes and finding sequence variations, but I think on the whole it has turned up a lot less than people had hoped," he says. "That indicates that the answer is someplace else."[7]

Regulome's pilot project is employing high-powered computers to sift through the junk DNA, hoping to find what he calls "regulatory regions." Maybe this genetic material holds defects that can lead to disease, Stamatoyannopoulos says.

Regulome's project is already gaining the attention of high-profiled investors such as Microsoft-cofounder and billionaire Paul Allen, who has funneled millions of dollars into the project.

Whether junk DNA's job is to regulate gene functions, as Regulome supposes, or has another yet-to-be seen purpose remains unclear. But today, most scientists believe the purpose will soon be revealed.

An interesting note: The human genome has a much bigger percentage of junk DNA (more than 50 percent) than most other organisms do. The roundworm has only 7 percent junk DNA, and the fruit fly, only 3 percent.

WHAT'S LEFT TO DO?

Scientists say the final sequence of the human genome was not the end, but a beginning. And there remains a long to-do list. Here's a sampling:

* Determine the exact locations and functions of all the genes.

* Discover how and by what means genes regulate other genes.

* Discover why some genes are expressed in some cells but not others.

* Determine the true function of junk DNA.

* Discover how gene expression coordinates with the making of proteins.

* Discover how to make predictions about how proteins fold.

* Figure out how to adequately predict gene function.

* Decode the proteome; figure out the actual protein content and function.

* Come up with predictions on disease susceptibility using single base DNA variations (single nucleotide polymorphisms, or SNPs) among humans.

We Show Up Late on the Family Tree: Really, Really Late

All that genetic overlap between mice, men, and countless other life-forms means we all have a single ancestor in common. Scientists say it was probably a single DNA or RNA molecule.

Calculating the amount of genetic overlap between life-forms tells us about our relatedness to another given being on a family tree—the more genetic similarities, the more related the two are. But it also means we can roughly tell when humans diverged from other earthly animal species. For instance, scientists believe that humans branched off from a common ancestor or a group of common ancestors between 150,000 and 300,000 years ago. We probably split from yeast about a billion years ago.[8]

Males Carry Most of the Mutations

The male mutation rate is roughly twice the female rate. That means that the male half of our species probably is responsible for most of the disease-causing mutations, but also for most of the mutation-related improvements.

In retrospect, that makes sense. The more cell divisions there are, the more chances for errors. Certain genetic sequences just may not copy correctly. The number of cell divisions required to make a sperm cell in a thirty-year-old man is about 400, because a man creates sperm throughout his lifetime. This explains why older men, with older sperm that have undergone even more mutations, father more children with genetic birth defects. A woman's eggs, however, require only about two dozen cell divisions, and those eggs are complete and set aside before the female is even born. (However, a March 2004 study published in *Nature* suggests that female mammals possibly can replace damaged eggs after birth.)

One Gene Codes for More Than One Protein

One of the oldest saws in biology—maybe you remember it from high school—is that each gene provides the code telling the cell how to build one protein. Not true. We are now beginning to understand that one gene might hold the formula for building multiple proteins, depending on where in the body the cell that houses it might be located.

"It turns out that a gene makes a message, but the message can be spliced up in different ways. And so a gene might make three proteins or four proteins, and then that protein can get modified," says Eric Lander of the Massachusetts Institute of Technology (MIT).[9]

So we have 30,000 genes, but 50,000 or more proteins in our body, scientists say. And maybe there are millions of proteins, once you consider all the tiny modifications that are possible. This may explain why humans only have a few more genes than flies, and even fewer than some other animals. The difference in complexity might not lie in the genes. It might be all about proteins.

A typical human gene knows how to make twice as many proteins as a fruit fly gene does. The proteins themselves are more intricate, too.

The Genome Is Lumpy

The image most people have of genes is that they are dotted evenly along the chromosomes, sort of like highway markers. Not so. The genes are located in clumps divided, in some cases, by vast areas of repeating stretches. You can look along a sequence of two million letters and not find a gene in sight.

The most populous chromosome is chromosome 1. It has more genes than any other chromosome: a total of 2,968. The most arid chromosome is the male Y chromosome, which only carries 231 genes.

> It's true you can't build the puzzle without knowing the pieces. Coming up with the human genome sequence wasn't a step; it was a leap, a huge leap. No doubt about it. But there are hundreds and hundreds of more questions it raises.
>
> Geneticist Mark Hughes

Size Doesn't Matter

It is not the size of the genome that matters, researchers have concluded. As we go up the scale from single-celled creatures, what increases is the number of control genes.

In other words, there are genes that do the real work and genes that control what those genes do. And there are genes that control the genes that control what the other genes do.

When you compare humans to, say, mice, you find that there are not as many new human genes that perform new human functions. This easily explains the vast similarities between mice and human DNA. What is new is the variety and sophistication of the control genes.

The Work Is Just Beginning

"It's true you can't build the puzzle without knowing the pieces," says Mark Hughes of the Genesis Genetics Institute in Detroit. "Coming up with the human genome sequence wasn't a step; it was a leap, a huge leap. No doubt about it. But there are hundreds and hundreds of more questions it raises."

Eric Lander of MIT's Whitehead Institute has widely promoted this metaphor: "This is basically a parts list. It's just a parts list. If you take an airplane, a Boeing 777, I think it has like 100,000 parts.

FIGURE 3-1. Francis Collins, head of the U. S. government's human genome research project.

If I gave you a parts list for the Boeing 777, in one sense you'd know a lot. You'd know 100,000 components that have got to be there, screws and wires and rudders and things like that. On the other hand, you wouldn't know how to put it together. And I bet you wouldn't know how it flies."

Figuring out the sequence is a long way from understanding what individual genes and the proteins they create actually do. Scientists don't even understand how proteins fold, not to mention where and when they do it. As Celera's Venter has pointed out, "The biggest danger is in overpromising. That's happened so many times over and over when there's a basic science advance like this."

But while there is a lot of hype surrounding drug companies and their ability to rapidly create drugs based on the new knowledge of the human genome sequence, "one thing is very clear," Venter says. "I spent ten years trying to find one gene. That now can be done in a fifteen-second computer search on our Web site and thousands of scientists [do] that today, saving ten years of research with that fifteen-second search."[10]

> *Think of it this way. Having this information available on the*
> *Internet for any scientist with a good idea . . . allows an*
> *empowering of all the brains of the planet to work together*
> *to try to understand what this book is telling us. To move*
> *into those medical advances we all dream of and deserve.*
>
> *Scientist Francis Collins*

The fact that the entire sequence is available on the Internet for anyone to search and take advantage of means genomic advancements are coming faster than would have been otherwise possible. "Think of it this way," says Francis Collins, director of the National Human Genome Research Institute. "Having this information available on the Internet for any scientist with a good idea . . . allows an empowering of all the brains of the planet to work together to try to understand what this book is telling us. To move into those medical advances we all dream of and deserve."[11]

Or, as MIT scientist Eric Lander noted, while a parts list isn't enough to know how a Boeing 777 flies, "of course, you'd be crazy not to start with the parts list."[12]

There are at least a dozen metaphors for the human genome sequence and what its true meaning is, but one thing is certain. The final decoding of the human genome in April 2003 wasn't the end of anything. It was only the beginning.

It's a Fact

Fact: The first letter in the Book of Life is a G. That is, the base guanine (G) is the first base on chromosome 1.

Fact: The most crowded chromosome is chromosome 1, with 2,968 genes. The least populated chromosome is the Y chromosome, with only 231 genes.

Fact: Most genes are about 3,000 bases long. The largest human gene—the dystrophin gene—at about 2.4 million bases, is on the X chromosome. Dystrophin is one of the key proteins involved in building strong muscle tissue. Boys born with a mutation in this gene end up suffering from the disorder known as Duchenne muscular dystrophy. Girls who inherit the mutation only carry the disorder, but don't suffer from it, since they also inherit an extra, healthy X chromosome from their father.

Fact: Scientists still don't know what more than 50 percent of genes do.

CHAPTER FOUR

THE DNA FILES

THE IDEA of criminal DNA fingerprinting first hit hard in the public consciousness in 1995, during the infamous O. J. Simpson trial.

The term DNA appears in the trial transcripts about 10,000 times. An expert for the prosecution even claimed that the chances of a drop of blood found on Simpson's shoe belonging to anyone other than his murdered wife, Nicole Brown Simpson, were one in twenty-one billion.

Despite the overwhelming scientific evidence on the table, Simpson was acquitted. His defense team successfully questioned the quality of the DNA samples, even suggesting that the blood was "planted."

Peter Neufeld, one of the attorneys on Simpson's "dream team," told me the defense team never tried to argue that DNA testing wasn't reliable, only that police had bungled the handling of the DNA evidence. "And the results are only as good as the integrity of the evidence before it gets to the lab." In court, Neufeld and his team argued that because the police department admitted mishandling the blood evidence, how could a juror have confidence in the case?

"The silver lining is that crime laboratories all over the United States didn't want to look like the Los Angeles police crime laboratories, which looked like bumbling idiots, so they've tried to clean up their own house," says Neufeld.

And, since 1995, DNA fingerprinting has since become easier and juries now understand it better. Crime labs once needed a fairly large sample of DNA from the blood, semen, or saliva left at a crime. That sample was easy to contaminate, and testing it could sometimes take weeks. Now, you only need a few cells to test, and coming up with results takes only about $50 and a couple of hours.

Neufeld and Barry Scheck, also a Simpson attorney, have started the Innocence Project, an effort based at the Benjamin N. Cardozo School of Law at Yeshiva University in New York City. The nonprofit's mission is to help wrongfully convicted inmates prove their innocence via DNA testing.

I'm so happy. This tells the world that I'm innocent.
Former prison inmate Kirk Bloodsworth, after DNA testing helped exonerate him on rape and murder charges

"What we realized as far back as 1989 is that [DNA evidence] is a much more robust technology than what they'd been using for thirty years. We always had a feeling that eyewitness identification and things like that weren't terribly reliable," says Neufeld.

Neufeld and Scheck decided to go back and see for themselves, reexamining convictions to see whether, in fact, police had snagged the right guy. In 1993, Kirk Bloodsworth, a Maryland fisherman, became the first person in the United States convicted in a death penalty case to be exonerated through DNA testing. The testing eliminated him as a source of semen stains on the girl's underpants and implicated a prison acquaintance of Bloodsworth as the real rapist and murderer. Bloodsworth served nine years in prison (two of them on death row) before a Maryland judge released him. The Maryland governor later pardoned him.[1] At this writing, the Innocence Project has helped exonerate 144 inmates, with more than a dozen of them formerly on death row. Hundreds more are waiting in the wings to get tested. It has also helped spawn what Neufeld calls a "national civil rights movement," a network of twenty-five innocence projects at law schools and universities around the country.

According to the Innocence Project, for every seven people sentenced to death in the United States, one is exonerated because he is found innocent. DNA fingerprinting is the gold standard used in determining that innocence.

The Gold Standard

Using DNA sequencing techniques, technicians are able to identify a set of "markers" that are different for every person, with the exception of identical twins.

British geneticist Sir Alec Jeffreys stumbled upon the advance of identifying people with a unique DNA signature that we now call the DNA "fingerprint." Within a year, Scotland Yard used the technology not only to exonerate a person accused of the murder of two girls, but also to find the individual who was guilty.

In a speech he made upon receiving the 1998 Australia Prize, Jeffreys says he remembers the exact "eureka" moment that he discovered a means to identify individual humans using their DNA. It was 9 a.m. on September 15, 1984.

"I thought, my God, what have we got here? It was so blindingly obvious," says Jeffreys, adding that he and his team at the University of Leicester were looking for genetic markers for very basic genetic analysis. "We [realized] we'd stumbled on a way of establishing a human's genetic identification. By the afternoon, we'd named our discovery DNA fingerprinting."[2]

In a nutshell, Jeffreys and his team were studying the gene for the protein myoglobin, a close relative of hemoglobin. But the scientists noticed that a big portion of the gene didn't code directly for the protein at all. Instead, it had "stutters," locations on the genome where certain bases repeated ten to fifteen times, over and over. To borrow an analogy often used in biology textbooks, if you think of the gene as saying "Mary had a little lamb," they noticed that in some cases the sentence read, "Mary had aaaaaaaaaaaaaaaaaa little lamb."

THE INS AND OUTS OF PCR

If you follow the DNA sciences, you'll eventually run into the term "PCR." The acronym stands for polymerase chain reaction, which is a fancy way to describe a simple way of making lots of copies of a DNA sample. That way, an investigator can take an incredibly tiny sample of genetic material—say, a couple of drops of saliva from the back of a stamp or a phone receiver—and make copies of it so there is enough material to test.

The process, developed by ex-surfer and now Nobel Laureate Kary Mullis, is simple. Essentially, you put your DNA sample—you only need a couple of cells—into the PCR machine, close the lid, and let it do its work. The PCR machine works by capitalizing on a central fact of DNA science—that the As attract Ts and the Gs attract Cs.

The system heats up the sample so that the two strands in the DNA molecule come apart. Then, it cools it down—which gives each strand time to build a complementary strand. You now have four strands where you once had two. The cycle is repeated again and again, until technicians have untold numbers of DNA strands where there were once just a few, and each strand is identical to the original molecule it started with.

The chemical that helps the split strands of DNA find new partners is a kind of enzyme known as a polymerase. The polymerase functions as a scissor, cutting the paired DNA strands in two. The PCR machine gets its polymerase from a type of bacteria that lives in hot springs—so it is able to survive at the near-boiling-point temperatures that you need in order to split apart the DNA.

After coming up with the theory behind PCR, Mullis had this observation: "I could make as much of a DNA sequence as I wanted, and I could make it on a fragment of a specific size that I could distinguish easily. Somehow, I thought, it had to be an illusion. Otherwise, it would change DNA chemistry forever. Otherwise, it would make me famous. It was too easy. Someone else would have done it and I would surely have heard of it. We would be doing it all the time. What was I failing to see?"[3]

Jeffreys and his team looked a little closer, and noticed something exciting. That same sequence of repeats—they called it a "hypervariable region"—occurred not just around the myoglobin gene on chromosome 22. It showed up all over the place. And when they printed out the patterns of where those repeats appeared, it turned out there was an extraordinary difference between how many times and where those hypervariable regions appeared on the gene from person to person.

By the afternoon of September 15, Jeffreys says his team members were pricking their own fingers, trying to see if they could create "evidence" identifying them from drops of blood smeared on tissues and glass. It worked: "It was a classic case of basic science coming up with a technology which could be applied to a problem in an unanticipated way," says Jeffreys.

"Half the bands from a child's DNA fingerprint come from its mother and half from its father," he adds. "In paternity testing, you take the child's banding pattern and that of the mother and the alleged father. The bands on the child's DNA fingerprint that are not from the mother must be inherited from the true father. And no two people have the same DNA fingerprint, other than identical twins."

> *I have no background in science. My partner, Barry Scheck, has no background in science. In fact, like a lot of other lawyers, it was the difficulty in comprehending chemistry that moved us to law school in the first place. But what happens is that you have a client whose life and liberty are at stake, and it forces you to learn certain disciplines.*
>
> Innocence Project cofounder Peter Neufeld, U.C. Berkeley's "Conversations with History" symposium, April 27, 2001

With his discovery, Jeffrey's place as one of the pioneers of biology was set. James Watson, one of the co-discoverers of the DNA double helix, told me that he considers Jeffreys's work to be the most important result of DNA technology so far. "When you talk about the [innocent] people on death row, well, you are talking literally about saving lives. What could be more important than that?"[4]

Saving Lives

Calvin Willis spent twenty-two years—almost half his life—in prison for a crime DNA evidence says he didn't commit

"I feel great about getting out of prison," Willis said in a phone interview with the *Los Angeles Times* after he was released, his life sentence with no possibility of parole commuted. "But to be honest, I am disappointed in the system—the unjustness of holding me here all this time. And I am not the only one who has suffered like this."[5]

Willis was convicted of the 1981 rape of a ten-year-old girl in Shreveport, Louisiana. He has always claimed he was innocent. The state convicted him after matching blood on boxer shorts left at the scene of the crime with Willis's blood type, which is a common one. But DNA investigators tested the blood again in 2003 and discovered that the male DNA found on the shorts definitely did not match that of Willis.

In September 2003, he became the 138th death row inmate in the United States to be exonerated by a DNA test.

"DNA aids the search for truth by exonerating the innocent. The criminal justice system is not infallible," says Janet Reno, the former U.S. Attorney General during President Bill Clinton's administration.[6]

Even Scheck and Neufeld, who have represented almost 70 percent of exonerees, were surprised at how many inmates turned out to be innocent.

"It's certainly counterintuitive," says Neufeld. "The DNA exonerations point out the system failure of criminal justice. So many of these people were identified on jailhouse snitch information, hair comparisons, bite mark comparisons, informants, and misidentification. And false confessions. As a public defender, I figured that if my client signed a confession it was basically the end of the case.

> *So many of these cases are literally wars.*
> *Barry Scheck, Innocence Project cofounder, PBS* Frontline *interview,*
> *October 31, 2000*

"But we see that sometimes these people are just broken down. They're told, just admit it, and we'll go easier on you. Or they may have a mental infirmity. But we now know, through the gold standard of DNA testing, that lots of people confess to crimes they didn't commit.

"The issue isn't that there was an unfair trial or an obscure mistake in the legal process. The people being exonerated by DNA fingerprinting are actually innocent—they did not commit the crimes for which they were incarcerated or, even, sentenced to death.

"This is total system failure of criminal justice," Neufeld says. "But things are starting to change."[7]

HOW DNA TESTING WORKS

There are many technologies for taking DNA fingerprints, but the two most accurate and commonly used are the methods called RFLP and PCR.

RFLP—short for restriction fragment length polymorphism—is the most accurate. It is also the most difficult test to do and the most expensive. That's because you need a fairly large sample of blood, semen, or skin, and it is far too easy to contaminate. RFLP is accurate to the degree of one to a billion or even better. Because of its accuracy, some states (California is one) only allow this method of DNA testing.

The PCR method—short for polymerase chain reaction—is not quite as accurate. But because PCR works as a kind of biological copying machine—you can take just a few cells and amplify that sample millions of times to get a lot of material to test—you don't need much evidence. And the method is fast. PCR works its magic in just a few hours, allowing forensics examiners to turn around their evidence overnight.

Illegal Search and Seizure?

Exonerating the innocent is only one side of DNA fingerprinting. Another side is the creation of databases of DNA fingerprints by the FBI, the Department of Defense, and law enforcement agencies. That is decidedly more controversial.

The FBI has been collecting DNA samples for its Combined DNA Index System (CODIS) from the police departments in all fifty states. The Department of Defense now collects DNA samples of everyone in the military. And there's a move afoot among U.S. law enforcement agencies to take DNA samples of every person arrested, not just those convicted of crimes. Police in Louisiana already do.

Many are concerned that in taking DNA, the most private possession a person has, the law is entering amendment-busting territory.

> *This is total system failure. Were not talking about . . . some procedural due process matter, some matter of unfairness in the way a trial was conducted. We're talking about people who are actually innocent. And that has to command our respect and attention and concern unlike any other kind of case.*
>
> *Barry Scheck, Innocence Project cofounder, 2000*

Philip Bereano, a professor of technology and public policy at the University of Washington and an outspoken member of the Council for Responsible Genetics, told me he finds that deeply disturbing. "I call it unlawful search and seizure, a violation of the fourth amendment."[8]

Benjamin Keehn, a Boston public defender, took the point to the American television audience on *The NewsHour* on PBS. "The state is saying, in effect, you may be a danger in the future because you were in the past, and therefore we need to register your DNA. That is a fundamentally different way than government has heretofore been permitted to treat its citizens.

"And if that theory prevails, it can be applied to any number of other potential classes or subclasses of our society, as to which an argument could be made that that subclass is at risk of committing crimes in the future," Keehn says. "If we are going to take DNA from prisoners because they are at risk, why shouldn't we take DNA from teenagers, from homeless people, from Catholic priests, from any subgroup of society that someone is able to make a statistical argument of being at risk [of committing crimes]?"[9]

To law enforcement's claim that DNA collection is necessary in order to solve crime, Bereano is resolute. "Of course, there are crimes that are being solved by the use of these databases. No civil libertarian is ever going to argue that violating civil liberties does not produce socially desirable results. After all, if the police were allowed to kick down doors without a search warrant, they'd find all kinds of things—people selling drugs, people beating their wives. They'd uncover crimes that they don't uncover now, but this is not the values the founders [of the United States] were after. [DNA databanking] is a radical departure from the balance [between civil rights and law enforcement] that the founders were after."[10]

The U.S. Court of Appeals for the Ninth Circuit Court looked hard at this issue in October 2003. In the case of *United States v. Kincade*, the court ruled that parolee Thomas Kincade did not have to submit a DNA sample to the FBI CODIS databank as it required. "We conclude that, as a matter of general Fourth Amendment law, forced blood extraction from parolees requires individual suspicion," wrote Judge Stephen Reinhardt in the opinion.

"Kincade's protest should serve more broadly as a warning about a technology that is popular, has expanded rapidly and, at times, irresponsibly, and whose most avid supporters have still not crafted adequate privacy protections or uniform standards for its use. On the issue of DNA databases, there are many unanswered questions and potential dangers," wrote Christine Rosen, a fellow at the Ethics and Public Policy Center in Washington, D.C., on the Web site of the *National Review*.

"Most Americans believe that the DNA at issue in these cases is just like a fingerprint—a harmless source of identification. But DNA is fundamentally different from a fingerprint—it is much more revelatory. This has led to misunderstanding about the benefits and dangers of DNA and DNA databases. The DNA stored in these databases is called, incorrectly, 'junk' DNA, because it is supposed to reveal only your unique genetic identity, not the details of your entire genome," added Rosen. "But junk DNA does not merely serve as a unique identifier; it can also reveal genetic predispositions for conditions such as Type I diabetes."[11]

> *The most satisfying application of our insights was Jeffreys's invention [of DNA testing]. It is exonerating people on death row; it is literally saving lives.*
> James Watson, co-discoverer of the double helix, 2003

Finally, there is concern about whether DNA carries an undeserved "imprimatur of legitimacy" in a court of law, says Troy Duster, previous head of the Human Genome Project's Ethic, Legal, and Social Implications (ELSI) board. Just because a person's DNA is found at the scene of a crime does not mean he is guilty, he told me. People leave DNA near crime scenes just by passing through.

Still more alarming, he goes on to say, is the increasing use of so-called "DNA dragnets," already common in England and Australia.

"After a crime is committed," Duster adds, "they ask every male in the neighborhood age 12 to 50 to contribute a sample, and those who say no are prime suspects. The police then follow them around, looking for samples they leave behind, a hair on a comb, a drop of saliva on a coffee cup or cigarette butt, a few specks of dandruff on the back of a chair."[12]

How can you interpret civil rights in an age of DNA technology? As courts wrestle with the implications, expect to hear more about this issue in months and years to come.

Who's Your Daddy?

DNA fingerprinting isn't just exonerating the innocent and, controversially, pointing at potential suspects. It's helping historians solve long-standing mysteries and issues of paternity.

Consider Charles Lindbergh. Most of us know him as the first man to fly solo across the Atlantic. But three Germans living in Munich today say they also knew him by another name: Carue Kent. They say he was their generous father, visiting them several times a year and supporting them with trust funds. But the time has come, they say, to correct the "father unknown" statement on their birth certificates and replace it with a name most everyone knows: Charles Lindbergh.

Dyrk Hesshaimer, Astrid Bouteuil, and David Hesshaimer claim that for the last seventeen years of his life, Lindbergh maintained a double existence, taking care of his American wife and children, as well as his German family.

"They look hauntingly familiar," Morgan Lindbergh, the aviator's legitimate grandson by marriage, told the Reuters news service after the siblings made the claim. It was the photos of the three Germans, he said, that pushed him to provide a DNA sample for testing.

The siblings say they don't want money, just the acknowledgement that Lindbergh is their father. "That is the most important thing we have to repeat," says Dyrk Hesshaimer.

> *When we're dead and gone, all that's left is a shell. I believe that if [Billy the Kid's mother] were alive, she would state, 'I'll gladly donate my DNA because I want you to prove that he is my son.'*
>
> Sheriff Gary Graves, De Baca County, Arizona, in "Billy the Kid's DNA Sparks Legal Showdown," MSNBC.com, November 12, 2003

The siblings, who were born between 1958 and 1967, have offered up a bundle of 112 letters allegedly penned by Lindbergh to their mother, Brigitte. They have childhood photos in which Lindbergh appears. They didn't realize that Kent could be Lindbergh until the late 1980s. But their mother swore them to secrecy, making them promise that they would keep the secret at least until both Lindbergh's wife, Anne Morrow Lindbergh, and she were deceased. Both women died in 2001. DNA tests done in 2003 confirmed what they already believed: All three of them were Lindbergh progeny.

EIGHT MYSTERIES SOLVED BY DNA

Where Is Columbus Buried?

The remains of Christopher Columbus have been moved so many times since his death in 1506 that no one has known for sure. Researchers are testing DNA from the body of his brother, Diego, against the remains in the Seville Cathedral tomb bearing his name. (The Dominican Republic also claims that Columbus's tomb is there.) It's known that Columbus did request burial on the Caribbean island now occupied by the Dominican Republic and Haiti. Historians long thought that the remains were moved to Cuba in 1795, and then Seville in 1898. But there is a Dominican artifact, an urn, with his name inscribed on it. Both sets of remains are being tested, so we may soon know the answer to this mystery.

Was Albert DeSalvo the Boston Strangler?

Probably not. In December 2001, DNA tests cast his guilt in doubt. DNA evidence found on the body of the Boston Strangler's last victim, nineteen-year-old Mary Sullivan, did not match DeSalvo. DeSalvo wasn't alive to see the results—he was murdered in his jail cell in 1973.

Did Sam Sheppard Kill His Wife?

He's been cleared. The famous physician who inspired the movie and TV series *The Fugitive* was accused of killing his pregnant wife in 1954. (His conviction was overturned in 1964 on a technicality.) Penniless and still professing his innocence, Sheppard died in 1970. DNA evidence exonerated him in 1997, when DNA from blood found out at the scene was found not to belong to either Sheppard or his wife, meaning a third person was at the scene of the crime.

Did Thomas Jefferson Father Children with His Slave, Sally Hemings?

In all probability, yes. Researchers matched Y chromosomes from one of Hemings's sons to a direct descendent of one of

Jefferson's paternal uncles. Because Y chromosomes pass from father to son, Jefferson is likely the father of at least one of Hemings's children. (The Jefferson family points out that one of Thomas Jefferson's relatives could've fathered the children, but historians consider that unlikely.)

Did Jesse James Die in 1882, or Did He Fake His Death?

He probably died. Scientists at George Washington University have matched DNA in teeth from a Missouri grave to that of descendants of James's sister. How do they know the teeth were from James and not one of his male relatives? They don't. Some claim that James was actually buried in a grave marked J. Frank Dalton.

Could the Romanovs Have Survived the Russian Revolution?

We now know most did not. With the help of a DNA sample from Prince Philip of England, scientists have concluded that the body found in a mass grave in the Ural Mountains in fact belongs to the murdered Czar Nicholas II. Using mitochondrial DNA testing, they confirmed that a woman in the grave was the mother of the three children also found. The bodies of the two youngest Romanovs, Alexis and Alexandra, have never been found.

Was Anna Anderson Really Anastasia?

No. Anna Anderson, who died in 1984, long claimed she was the Princess Anastasia Romanov, daughter of Czar Nicholas II. Many believed her, but no more. After her death, DNA testing proved that in fact Anna Anderson was Franzisca Schonkowska, a Polish factory worker reported missing for years.

Did the Last Dauphin Escape?

No. DNA samples support claims made by French revolutionaries that Louis Charles, the ten-year-old son of Louis XIV, died in jail. Mitochondrial DNA samples from the boy's mummified heart match DNA in the hair of his mother, Marie Antoinette.

Where the Bodies Are Buried

It's become a challenge to keep up with the historical mysteries that DNA technology is helping to solve. But the puzzle that DNA testing may be best at solving involves setting the record straight on who is buried where.

In the case of several historical legends, that question is more relevant than ever. Who is actually buried under Billy the Kid's tombstone in Fort Sumner, New Mexico? As legend and history have it, The Kid (aka William H. Bonney, Kid Antrim, and Henry McCarty) shot and killed two deputies and was in turn shot and killed by Sheriff Pat Garrett in July 1881.

But there's reason to doubt that claim.

> *Pat, you son-of-a-bitch, they told me there was a hundred Texans here from the Canadian River! If I'd a-known there wasn't no more than this, you'd never have got me!*
>
> Billy the Kid to Pat Garrett in 1881, immediately after stepping out of the rock house at Stinking Springs and surrendering to Sheriff Garrett's posse

In 1950, an elderly Texan by the name of "Brushy Bill" Roberts claimed that he was Billy the Kid, and that Sheriff Pat Garrett shot someone else that day and covered up the lie. (There never was a formal investigation.) If true, it's an important historical footnote that would turn Pat Garrett from a folk hero into a murderer.

"We want to get to the bottom of it," says New Mexico Governor Bill Richardson. He supports exhuming the grave of Billy the Kid's mother in order to test her DNA against that of Brushy Bill. If they match, the governor will likely pardon The Kid.

People are curious, Richardson explained to a Voice of America reporter at the height of the controversy. "When they're confronted with . . . Billy the Kid, who's a western legend, people might say, 'Hey, he might not have been as bad as everyone says he was; in fact he was supposed to be pardoned and he wasn't,' and maybe he wasn't killed by Pat Garrett. Maybe he didn't kill those lawmen. So let's look into it. All we're doing is with science, with historians, and with national labs that have technology."[13]

And Who Are We, Anyway?

Until recently, the only way anthropologists could study our evolutionary roots was to dig up skulls and investigate remains. So they're understandably excited to have DNA printing as the newest tool in their arsenal.

Svante Paabo of the Max Planck Institute for Evolutionary Anthropology in Leipzig, Germany, is putting it to good use. He's developing a side-by-side comparison of humans and our closest relatives. It is the first genome-wide comparison of human beings and chimpanzees, he says. Ideally, it will help scientists figure out when certain human features appeared during the course of evolution.

> *I thought, my God, what have we got here? It was so blindingly obvious . . . we [realized] we'd stumbled on a way of establishing a human's genetic identification. By the afternoon, we'd named our discovery DNA fingerprinting.*
>
> Sir Alec Jeffreys, inventor of the DNA fingerprint

Paabo told me about one comparative DNA study, in which he discovered that humans and chimps evolved more recently from a common ancestor than did chimps and gorillas. In other words, we are more like chimps than gorillas are. He also discovered that there is ten times greater variation among any two random orangutans as there is among any two unrelated people.

Paabo says the team also discovered twenty-one changes related to hearing between chimps and humans. "It is fascinating to think this might have something to do with language," says Paabo. "One could imagine, for example, that language has caused special new requirements on hearing to appear in humans." Because scientists have now pinpointed the genetic source of differences between humans and chimps, they will certainly be looking at ape hearing more closely in the future, he adds.[14] That will help scientists better understand hearing in humans.

And then there's your mtDNA—your mitochondrial DNA. Until this chapter, I've been talking about your genes as if they're all a combination between your father's and your mother's. And for

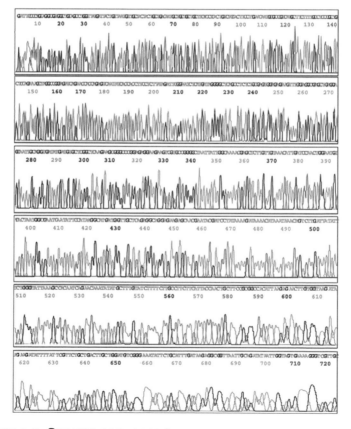

FIGURE 4-1. Genome sequence trace.

twenty-two pairs of your chromosomes—your so-called autosomal chromosomes—that's exactly the case.

But there are two exceptions to the rule: One is the Y chromosome brought to boys by their fathers, and the other is the mitochrondrial DNA handed down exclusively through the maternal line. That latter kind of DNA—the mtDNA—isn't located on the twenty-three pairs of chromosomes we've been talking about so far. Rather, it's outside the nucleus, located inside a little cell organ called the mitochondria, which is responsible for regulating cell energy. The DNA controlling it is located on a circular strand inside.

Because the DNA in the mitochrondria doesn't recombine with the father's DNA every time a couple has children, it stays pure. That means, the mitochrondrial DNA you have in your cells is exactly the same as the mitochondria in your mother's cells, your mother's mother's cells, and so on. It is a perfect line of descent. That makes it theoretically possible to trace back the DNA in all our mitochrondria to a handful of original females.

According to University of California at Irvine mitochondrial geneticist Douglas Wallace, there are eighteen such daughters, each with a separate and distinct mitochondrial DNA arrangement indicating from what region of the world they radiated. There are ancestral Asian, American Indian, African, and European lineages. After he compared the mitochondrial DNA sequences of a group of individuals from the Kalahari Desert in southern Africa, Wallace found their DNA to be among the most ancient at 145,000 years. (DNA itself, of course, is much older than our species.)[15] This supports the idea that all humans arose from Africa, and that "ancestral Eve" was likely a very hardy black woman. To survive in ancient times, she would have had to be hardy.

FACING DESTINY

LIKE SO MANY little girls, Colorado second-grader Molly Nash wears bangs, plays soccer, and likes school—reading especially. But she didn't always seem so average.

Her mother, Lisa, knew something was wrong the minute she delivered Molly. "They didn't let me see her. They just whisked her off, and all I heard in the background was everybody saying, 'There're abnormalities of the hands and the forearms.'"[1] It turned out Molly had a rare genetic disease called Fanconi's anemia, and both her parents were unknowing carriers. Because of the genetic mutation she'd inherited, bone marrow cells responsible for building her white blood cells were failing. She was born with no thumbs, a perforated heart, and deafness in one ear.

Doctors said she was unlikely to reach the age of six. But a controversial treatment called preimplantation genetic diagnosis (PGD) changed all that. Using it, doctors were able to help Molly's parents conceive a second child that didn't have the genetic defect. That child, Molly's little brother Adam, donated the umbilical cord blood containing the cells Molly needed for a matching bone marrow transplant. In an interview with ABC's *20/20*, Lisa says Molly understood.

"Yes, she knew that, and that's what she told everybody who came to our house to kiss us good-bye. 'I'm going to Minnesota, and my brother's blood will make me healthy.' And we put him in her lap. They hooked up this bag with this . . . liquid gold. It was very peaceful. It was very calm. We all held each other, and forty-five minutes later, her new life began."[2]

Molly's doctors are quick to point out that Adam is not a designer baby, which is a common misunderstanding. In fact, when I interviewed Mark Hughes, a geneticist with Genesis Genetics Institute and the Wayne State molecular biologist widely considered to be the pioneer of this technology, he expected me to make the criticism—and the first thing he said was defensive. "People have children for all kinds of reasons: money, power, companionship, to save a marriage, to work on the farm, tax reasons. So what's the matter with this? This technology brings the miraculous power to cure a sister."[3]

The actual PGD procedure is fairly straightforward. First, a couple uses the same method used in in vitro fertilization (IVF) to generate a large number of embryos. After enduring four IVF cycles, Lisa and her husband, Jack, generated thirty embryos. Then, geneticists examine each embryo to find the genetic problem—in this case, the mutation that causes Fanconi's anemia. Five of the embryos were free of the mutation and also turned out to be a tissue-match for Molly. Four of the five embryos refused to take hold in Lisa's womb. The fifth did take hold, and Adam was born nine months later.

More than 1,000 children have been born through PGD, says Hughes. He is currently one of the most sought-after fertility specialists in the world.

"These babies are born to couples whose children are at risk for serious genetic diseases, couples that don't want to just throw the genetic dice," he says. "This is a technology that assures a healthy baby."[4]

If we could honestly promise young couples that we knew how to give them offspring with superior character, why should we assume they would decline? If scientists find ways to greatly improve human capabilities, there will be no stopping the public from happily seizing them.

James Watson, co-discoverer of the double helix

The technology has detractors, however. According to the press reports, the Vatican contacted Molly Nash's doctor after news of her brother's birth reached Vatican City. The Catholic Church has similar objections to PGD as it does to in vitro fertilization, because extra embryos are created and some are destroyed. Still others have concerns that geneticists may soon give parents the ability to screen for traits other than predisposition for serious diseases. Traits such as gender.

"That's a valid concern," says Hughes. "Gender isn't a disease. No one should test for gender." But Hughes says he can test for 217 gene mutations covering about eighty-six inherited diseases. It's a list that keeps growing.

That's precisely what ethicists worry about. They contend that soon there will be a blurry line between screening embryos for life-threatening childhood diseases such as Fanconi's anemia and genetic predispositions to such diseases as cancer, Alzheimer's, and Huntington's chorea, which don't generally appear until a person has already lived a fairly long and healthy life.

Matt and Denise Rominger are the first couple known to have screened their embryos for Huntington's chorea, a fatal genetic disease resulting from excessive repeating of the triplet CAG on chromosome 4. Matt's mother died of the disease, as have several other relatives. In 1992, he was tested and discovered that he too had the mutation. And any child he fathered would also have a 50 percent chance of having it. PGD seemed a natural way of having children who didn't share the risk.

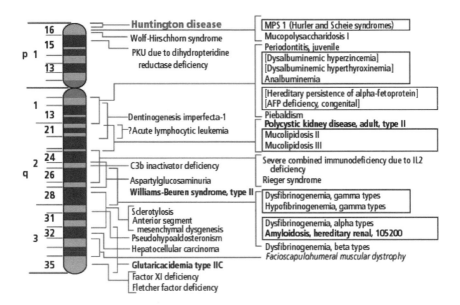

FIGURE 5-1. Chromosome 4 and potential locations for mutated genes.

Matt and Denise went ahead with the expensive, time-consuming procedure. He still shows no signs of the disease—and his twin daughters, Austin and Hannah, are vibrant elementary schoolers.

But is there a slippery slope to worry about? If parents can select for such undesirable mutations leading to birth defects and debilitating illness, why not select for intelligence or athleticism or green eyes or long legs? It's science fiction now, but what happens if and when it becomes possible?

There is no doubt someone would like to try it. As James Watson noted in his book, *A Passion for DNA:* "If we could honestly promise young couples that we knew how to give them offspring with superior character, why should we assume they would decline? . . .

If scientists find ways to greatly improve human capabilities, there will be no stopping the public from happily seizing them."[5]

Mark Hughes says the widely held public fear of "designer babies" isn't realistic. For one thing, IVF is expensive and uncomfortable. "There are easier ways to get pregnant," he says. And then there are probabilities making it highly unlikely anyway. Almost all traits are a combination of several genes, not just one. And they must be present in both parents for them to pass a trait onto their child.

"Even with IVF and PGD," says Hughes, "Dr. Ruth could not have Brooke Shields, and Danny DeVito could not have Arnold Schwarzenegger, except perhaps in the movies."[6]

Why is this? Hughes puts it this way: "Suppose a couple, who didn't appreciate well the wonders of creating a child for its own sake, wished to have an offspring that was as brilliant as Albert Einstein. And suppose that in the thousands of genes that are expressed in the human brain, there are only six that are involved in intelligence—and this is surely an underestimate." For the Einsteinian baby to be born, he explains, the man and the woman would both together have to carry the six intelligence genes. In other words, they can't give to a child what they don't have.

> *Ninety-nine percent of people don't have an inkling about how fast this revolution is coming.*
> Affymetrix founder Steve Fodor, quoted in Matt Ridley's Genome
> (New York: HarperCollins, 1999), p. 258

Hughes continues: "These six genes are like cards in an enormous genomic deck, and they can be shuffled and distributed in almost endless combinations. This is why couples will marvel at how their children are totally different from each other, yet they came from the same parents. Assume that the woman and the man actually do have these six 'intelligence genes'—say, three each—and that they are inherited in a 'dominant' [50–50] manner, such that each gene's presence is needed for Einstein brilliance."

In this scenario, the woman would have a 50 percent chance of giving each of the three intelligence genes through one of her eggs. Her chances of having an egg that would contain all three would be one in eight. And the odds of the couple actually producing an embryo with all the genes required is one in sixty-four.

This is where you can see the difficulty. As IVF experts are quick to point out, the best fertility centers in the world only succeed in producing eight to ten eggs, only a few of which actually survive after implantation. "It would be unethical and medically irresponsible to give the woman enough hormones to cause her ovaries to overproduce in this way. Even if IVF and PGD technologies improved a thousand percent from today's standards, biology will prevent such an abuse of this procedure. In this example, all of the assumptions are simplified, and it is likely that there are many more genes required than just six," Hughes says.

"Parenthetically," he adds, "I would argue that any couple who wanted to perform PGD for this purpose, de facto, does not have *any* of these intelligence genes, but that's another subject."

Taking Your Chances

Genetic testing of embryos is a hot-button issue, but genetic testing for prospective parents and adults at risk for hereditary diseases is a wave that has already struck.

Many pregnant women now undergo genetic tests. So do most newborns. And a growing number of adults are getting the opportunity to find out what risks lie dormant in their DNA.

Testing costs between $100 and $2,500, depending on the type of test. And there are already more than 950 genetic tests available in the doctor's toolbox, tests that can identify a genetic risk for a wide range of diseases. It is the single largest application of genomic knowledge to date. Couples use genetic tests for preconception and newborn screening. There is carrier screening to help couples find out if they both carry a copy for a disease that requires two copies for it to be expressed. And, of course, adults who learn they have a parent or other close family members with a disease can choose to get testing to estimate their risk.

TWO KINDS OF MUTATIONS

Some mutations are hereditary. That means you inherited it from one or both parents, and it is present in the DNA in all your cells. Other mutations are acquired mutations, mistakes that develop over time because of your lifestyle, exposure to toxins, infections, and so on.

Mutations happen. Usually, a cell knows how to fix a mutation before its daughter cells inherit it. But not always. Sometimes DNA repair efforts don't work—or don't work as well. That happens as we age. Mistakes accumulate, and age-related problems can develop.

Marina, a young woman in Italy, falls in the latter category. One day, while she was talking to neighbors, she grabbed at her chest and fell to the ground. It appeared to be a heart attack, but it wasn't. And when she got to the hospital, doctors told her she'd be fine. That's because they'd known what was coming.

Marina suffers from a rare inherited heart condition called stress-induced polymorphic ventricular tachycardia. She discovered it through a genetic test after two of her sisters died of the condition. "Her risk of having a heart attack was high," Silvia Priori, an Italian cardiologist who treated Marina, told *Scientific American* magazine. So before any symptoms appeared, doctors implanted an automatic heart defibrillator in Marina's chest.

Whenever her heart fibrillates, as it did that day, the device can get it working again. Knowing her 70 percent risk of a heart attack at any time, waiting for an emergency vehicle to arrive would be just too dangerous.[7]

This was about power and drawing lines in the sand, and whether employers secretly, or by coercion, can force employees to divulge their genetic secrets.

Harry Zanville, lead counsel for Brotherhood of Maintenance of Way Employees, quoted in Kristen Philipkoski's "Genetic Testing Case Settled," Wired News, April 10, 2001

Of course, Marina's disease is a rare one. Only a few genetic tests are used in life or death situations such as hers. More and more frequently, however, genetic tests are being used to help people assess their risks so they can change their lifestyles and take proper precautions when necessary.

TESTING, TESTING

Below is a list of some of the commonly available genetic tests. Some of the tests, such as the Alzheimer's screening, only reveal whether you have an increased risk for a disease. Others reveal with virtual certainty that the disease will strike. Your genetic counselor will give you the odds and risk factors before testing.

Genetic Test	Disease/Symptoms
Adult polycystic kidney disease (APKD)	Kidney failure and liver disease
Alpha-1-antitrypsin deficiency (AAT)	Emphysema and liver disease
Amyotrophic lateral sclerosis (ALS)	Lou Gehrig's disease; progressive motor function loss leading to paralysis and death
Alzheimer's disease (APOE)	Late-onset variety of senile dementia

Ataxia telangiectasia (AT)	Progressive brain disorder resulting in loss of muscle control and cancers
Charcot-Marie-Tooth (CMT)	Loss of feeling in ends of limbs
Congenital adrenal hyperplasia (CAH)	Loss of feeling in ends of limbs and genitalia and male pseudo-hermaphroditism
Cystic fibrosis (CF)	Disease of lung and pancreas resulting in thick mucous accumulations and chronic infections
Duchenne muscular dystrophy/Becker muscular dystrophy (DMD)	Severe to mild muscle wasting, deterioration, weakness
Dystonia (DYT)	Muscle rigidity, repetitive twisting movements
Fanconi's anemia, group C (FA)	Anemia, leukemia, skeletal deformities
Factor V-Leiden (FVL)	Blood-clotting disorder
Fragile X syndrome (FRAX)	Leading cause of inherited mental retardation
Gaucher disease (GD)	Enlarged liver and spleen, bone degeneration
Hemophilia A and B (HEMA and HEMB)	Bleeding disorders
Hereditary hemochromatosis (HFE)	Excess iron storage disorder
Hereditary nonpolyposis colon cancer (CA)	Early-onset tumors of colon and sometimes other organs
Huntington's disease (HD)	Usually midlife onset; progressive, lethal, degenerative neurological disease

Inherited breast and ovarian cancer (BRCA 1 and 2)	Early-onset tumors of breasts and ovaries
Myotonic dystrophy (MD)	Progressive muscle weakness; most common form of adult muscular dystrophy
Neurofibromatosis type 1 (NF1)	Multiple benign nervous system tumors that can be disfiguring; cancers
Phenylketonuria (PKU)	Progressive mental retardation due to missing enzyme; correctable by diet
Prader-Willi/Angelman syndromes (PW/A)	Decreased motor skills, cognitive impairment, early death
Sickle cell disease	Blood cell disorder; chronic pain, and infections
Spinal muscular atrophy (SMA)	Severe, usually lethal, progressive muscle-wasting disorder in children
Spinocerebellar ataxia, type 1 (SCA1)	Involuntary muscle movements, reflex disorders, explosive speech
Tay-Sachs disease (TS)	Fatal neurological disease of early childhood; seizures, paralysis
Thalassemias (THAL)	Anemias; reduced red blood cell levels

Source: The Human Genome Project. Available online at http://www.ornl.gov/sci/techresources/Human_Genome/medicine/genetest.shtml.

The Dawn of Predictive Medicine

Human Genome Project leader Francis Collins sums it up best: "In ten years, we should be able to make predictions for you and me for what conditions we're most likely to be at risk for, and that in itself would allow us to practice some preventive medicine strategies based on our own individualized risks. Give us twenty years, and I think you won't recognize medicine in the way the therapies are developed and applied."[8]

Leroy Hood, president of the Institute of Systems Biology in Seattle, told me that he's banking on it. "I think in the next ten years, we are going to make enormous progress in terms of very early diagnostics, predictive tools. If you want to deal with cancer most effectively over the next five years, you want the ability to call it cancer—to diagnose it—really early."[9]

> *In ten years, we should be able to make predictions for you and me for what conditions we're most likely to be at risk for, and that in itself would allow us to practice some preventive medicine strategies based on our own individualized risks. Give us twenty years, and I think you won't recognize medicine in the way the therapies are developed and applied.*
>
> *Francis Collins, Human Genome Project leader*

In some cases, like cancer or Marina's heart condition, being able to predict what you're at risk for can save your life. The genetic test for the rare hereditary disease hemochromatosis is another example. One of the most common genetic disorders in the United States, one in eight to twelve people carry one copy of the gene responsible for hemochromatosis, which is also known as "iron disease." And about five in 1,000 have the two copies necessary to get the disorder. Basically, it causes the accumulation of iron throughout your body. Without treatment, it can result in liver damage, impotence, and diabetes. But bloodletting on a regular basis can entirely prevent the worst consequences of this ailment.

And consider the tests now available for various cancers. BRCA1 and BRCA2 gene mutations lead to a dramatically increased risk of breast and ovarian cancer, and mutated versions of the MLH1 and

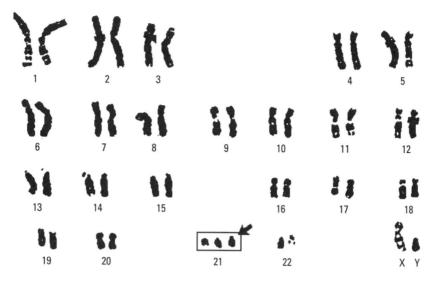

FIGURE 5-2. What happens when someone has Down's Syndrome—three Chromosome 21s.

MSH2 genes increase the risk for colon cancer. (Tests are now available to discover the presence of both types of mutations.) In both cases, affected people have options—they can change their lifestyles, get regular mammograms or colonoscopies, or even elect to have their breasts or colons removed altogether.

But what if a test exists for a disorder, but no treatment exists? There is a variant of the so-called APOE gene that leads to a significantly increased risk of Alzheimer's disease. Then again, more than 70 percent of people who have the variant APOE never get the disease. Should you be tested or not? According to experts in the United States and Europe, the answer is no. Because there is nothing you can do for an increased risk, they recommend that doctors only use the test to confirm a diagnosis, not to predict the likelihood of the disease.[10]

"Prediction without cure is anathema to medicine," says Hood.

Nancy Wexler, the neurologist who helped find the gene for Huntington's chorea, tells a story to make this point. There is no cure for Huntington's. Patients who inherit the gene are bound to die a slow, horrible death from this wasting neurological disease. In an essay, Wexler recalled how, as she tested hundreds of people in Venezuela for the illness, she was most preoccupied by the thought of how as-yet-unaffected people would react to the news. These are the people "who capture our imagination and concern."

> *Scientists and parents now have the power and the preroga-*
> *tive to decide what's a good enough inheritance to get not only*
> *into Harvard but also into life. But what are the new standards*
> *going to be, and where are they going to come from? Who has*
> *the wisdom to say these "improvements" are going to make us*
> *better human beings? We really are at a crossroads.*
>
> Bioethicist Leon Kass, as quoted by Rick Weiss in "Building a New
> Child," The Washington Post, *June 30, 2001*

Says Wexler, "As there are only a few such people, less than 100 people, we know very little about how this new 'presymptomatic group' will react to the bad news. One man pointed to the bridge over the lake near where we were meeting and said succinctly, 'If you tell me I am going to have this disease and I do not have someone to talk to about this, I am going to run to the nearest bridge and jump off!'

"We were unable to tell people when the disease will start: We can just say that they most likely have the gene. In follow-up interviews some time after testing, people who have tested positive were asked if they think they will develop the disease; some reply, 'I don't think so, because God will cure me, or science will cure me, or the test was wrong.' It is traumatizing to be totally healthy and know with almost 100 percent certainty that Huntington's disease is in your future," says Wexler.[11]

A LONG ROAD TO A CURE

Researchers have spent years looking for genes related to ovarian cancer, a deadly killer of women. Because it is so difficult to diagnose, coming up with a test is key. Ovarian cancer is often well into the serious stages before a doctor and a patient discover it.

In June 2003, researchers at Cancer Research U.K. in Edinburgh, Scotland announced a major victory. They discovered a gene called OPCML, which they found to be inactivated, or turned off, in more than 90 percent of the ovarian tumors they examined. When they turned it back on in ovarian cancer cells in the lab, cell growth halted.

Scientists at the lab say they believe that OPCML codes for a protein on ovarian cells that causes the cells to be "sticky" and adhere to one another. When the gene is inactivated, the cells don't stick together and unrestrained cell growth results. A potential drug that might result from this discovery would probably mimic OPCML's function.

"It is always heartening to make headway when investigating a cancer, like ovarian cancer, which is difficult to treat entirely successfully unless caught early. This work still has a long way to go in the laboratory before patients could benefit, but results so far are promising," says John Toy, Cancer Research U.K.'s medical director.

And that's always the rub, isn't it? While genomic discoveries are happening faster than ever, taking a discovery and getting it to a point where therapies and tests are possible takes years.[12]

Conversely, Wexler points out that people who receive extensive genetic counseling—people who truly understand the situation—do well. Genetic illiteracy, though, is a real enemy. People often don't understand that if you carry the gene for a dominant disorder, such as Huntington's or polycystic kidney disease, you will inevitably get the disease. However, carrying the gene for a recessive disorder (e.g., cystic fibrosis, Tay-Sachs) puts your children at risk only if your spouse carries it, too.

> *There's no disease, except some cases of trauma, that doesn't have hereditary contributions. There's not one example I know of.*
>
> Francis Collins, Human Genome Project leader

"How do we explain technically complex and emotionally charged information to ordinary people, many of whom never heard of DNA and barely [know] of genes, who have hardly a clue about probability, and whose science education never equipped them to make choices regarding these matters?" Wexler asks.[13]

These are questions that doctors and ethicists alike grapple with. Yes, gene tests are arriving far more quickly than associated therapies. Therapeutics will likely catch up in twenty or thirty years, predict Hood, Collins, and others. But that's a long time.

And, even stickier, genetic tests are becoming available more quickly than the public is learning about genetics in general.

A GENE THAT COULD GIVE YOU A HEART ATTACK

Are you more susceptible than someone else for suffering a heart attack or stroke?

Some scientists say that you could be if you have a less common variant of a gene called ALOX5. The gene, researchers from the University of Southern California and the University of California say, seems to be linked to atherosclerosis, or clogging of the arteries.

Scientists who studied 470 utility workers over the course of eighteen months discovered that those with a less common

variant of the gene had thicker artery walls, a typical atherosclerosis symptom. Diet, they found, played an important
role in reducing the increased susceptibility the gene causes.

"The adverse effect of this gene is increased by dietary
intake of certain n-6 polyunsaturated fats," says USC
researcher James Dwyer, "while the adverse effect is
blocked by intake of fish oils containing n-3 polyunsaturated fatty acids."[14]

Scientists think that the function of the ALOX5 gene variant could be to convert bad fats—such as those found in oil
and eggs—into artery-clogging molecules. Conversely, a
diet rich in so-called good fats, such as those found in fatty
fish, might reduce the damage.[15]

*Is it not interesting to note that medicine was, in its history,
first of all curative, then preventive, and finally predictive,
whereas today the order is reversed: initially predictive,
then preventive, and finally, only in desperation, curative?*

Nobel Laureate Jean Dausset, in the Journal of Biomedicine and
Biotechnology 1 (1), 2001, pp. 1–?.

A Rabbi's Story

The orthodox Jewish community in the United States stands as a
powerful example of what can happen when a community comes
together to address a dread genetic disease.

That disease is Tay-Sachs, a fatal neurological disorder caused by
gene mutations on chromosome 13. Like sickle cell anemia and cystic fibrosis, it is a recessive disease, meaning that a baby needs to
inherit the flawed version of the genes from both parents to get the
disease. If that happens, the baby is doomed. At about six months of
age, the baby starts falling behind on development milestones, by
eighteen months, she starts going blind and having convulsions.

Eventually, she becomes completely unresponsive. Tay-Sachs children only rarely live past the age of three.

One in about 300 people is believed to carry a faulty gene putting them at risk for this disease, but the rate is even higher among Jews of Eastern European, or Ashkenazi, descent. In that community, the odds are closer to one in twenty-five.

"After I lost my fourth child to Tay-Sachs and went through, together with my wife, all that suffering, I decided that something has to be done to stop this happening again in this community," says Rabbi Joseph Ekstein of Brooklyn, New York. In the 1980s, he helped put together a local effort, Chevra Dor Yeshorim (or the Association of an Upright Generation), that encourages young people to get screened for the disease before they marry. The organization funds the screening through contributions and public grants. So far, it has tested upwards of 90,000 people for this gene.

> What's clear is that as new pieces of technology become available, there's a debate that's not really a scientific debate. It needs to be out in the broader community about what's acceptable and what's not. The community says genetic testing that leads to family planning is okay in some cases. But for a disease thirty years out, does that change? You need lots of people involved in this debate before you come to a conclusion about what society wants to happen.
>
> Dr. William Thies, vice president of the Alzheimer's Association, quoted by Denise Grady in "Genes, Embryos, and Ethics," The New York Times, March 3, 2002, section 4, p. 6

Rabbi Ekstein says that screening has identified only a few hundred at-risk couples from the thousands it screens. As Orthodox Jews in general oppose abortion, he advises the couples to break up immediately rather than risk having Tay-Sachs affected children. "It's not easy, we know. But on the other hand, how can you compare the pain of seeing a child passing away in your own hands? How can you compare one with the other?"[16]

Looking Out for the Little People

But what about disease and disorders that are not fatal—disorders that are common enough to have their sufferers create thriving cultures around them? The blind and the deaf come to mind. So do the self-proclaimed "little people," the tiny men and women who live with dwarfism.

"The concept of a dwarf community is illustrated in part by the dozens of dwarf children from all over the world adopted by [adult dwarves] . . . there is a common feeling of self-acceptance, pride, and community," says Ruth Ricker, a past president of the genetic support group, Little People of America. The group boasts a membership of more than 5,000 individuals.

> *Even with [this technology], Dr. Ruth could not have Brooke Shields, and Danny DeVito could not have Arnold Schwarzenegger, except perhaps in the movies.*
>
> Mark Hughes, PGD pioneer, 2003

Ricker has been busy articulating a fear growing common among people born with disabilities. Now that prospective parents are increasingly able to screen out for various disorders, they may choose not to have children or to terminate pregnancies when abnormalities come to light. For Ricker—who, like many dwarves, suffers from a chromosome 4 mutation called achondroplasia—that hits close to home.

Ricker appealed to the public on PBS's *The NewsHour*: "We could see dramatically fewer dwarf children being born to average-size parents, and pressure on parents of all sizes to screen for and prevent the birth of what we would call healthy dwarf babies, kids that would grow up to be like me," she said. "We're still getting a feel for this dilemma, just as we have a first generation of us where a majority of us have had these opportunities, that we're now presented with the prospect that we may be gradually eliminated."[17]

> *A person taking a genetic test makes a terrific gain-loss calculation. The gain, obviously, is to learn that you do not have the genes for Alzheimer's, cystic fibrosis, Huntington's, or any number of other diseases. The loss is to learn that you do. Is learning the good news worth risking hearing the bad?*
>
> Gene hunter Nancy Wexler, "Clairvoyance and Caution," in The Code of Codes, *Kevles and Hood eds., (Harvard University Press, 1992)*

What it comes down to, ethicists say, is whether you consider dwarfism—or hereditary deafness or blindness, for instance—a disease. Through genetic testing, will we decide to discard people who seem to be too much trouble? At some point, will the only babies being born be the ones who fit in a narrow, predefined standard of what's normal and acceptable?

It would not surprise most ethicists when, in the relatively near future, an insurer hands prospective parents a list of a few hundred disorders to screen for during amniocentesis, saying that it will not cover them if the parents elect to give birth to an affected baby.

Insurance Companies

No discussion about genetic testing would be complete without mentioning insurance companies and the very real pall they already cast over many decisions to get tested. Maine Senator Olympia Snowe says that in one case, more than 30 percent of women refused a genetic test for breast cancer because of fear that a health insurance company would discriminate against them. Americans definitely shouldn't have to choose between being able to find out about their genetic profile and keeping their insurance, she says.

Snowe was inspired to introduce her so-called Genetic Nondiscrimination Act of 2003 after a letter from a constituent, Bonnie Lee Tucker of Hampden, Maine. She was diagnosed with breast cancer in 1989 and 1990. The disease had struck her mother and nine close relatives, too.

> *The real ethical dilemma that's going to confront us in the future—is that this technology is very expensive. People with money are going to be able to give their genetic enhancements to their children, and people without money are not going to be able to afford it.*
>
> *Ethicist Lee Silver*

A simple blood test will detect whether Tucker's twenty-five-year-old daughter has the BRCA1 or BRCA2 mutations, but she doesn't want her to have the test for fear her daughter will lose her insurance at such an early age.

The Genetic Nondiscrimination Act of 2003 would make people feel safer, no question. The legislation passed the Senate in October 2003 by unanimous vote, 95 to 0, and has yet to make it to the House. Specifically, it prohibits an employer from using genetic information in hiring, firing, job assigning, promoting, or making other employment decisions. Companies are also banned from collecting genetic information about employees or their families unless special criteria are met, such as when, for instance, the employer is worried about monitoring toxic materials' effect on workers.

More to the point, the bill keeps insurance companies from requiring genetic data before enrolling someone in a plan, and prevents them from using it to turn down customers or change or set rates.

"I hope that with this bill my daughter can be free of worries to be tested, so that she can go on with her life . . . . These companies are not going to save money on my daughter," she says.[18]

At the Senate hearings, insurance companies spoke out against the bill, calling it "unwise." There is already adequate protection for customers, testified Donald Young, president of the Health Insurance Association of America. "Imposing restrictions beyond those already in place could hurt the very people they are intended to help," he says, "by limiting the ability of insurers to appropriately and fairly set premiums."[19]

Real or perceived, the threat of insurance companies penalizing people for genetic knowledge could throw a wrench into the so-called genomic revolution, says Francis Collins, who comes down on Senator Snowe's side of the argument. Genetic discrimination, he testified, "is an area that could cause this wonderful revolution fueled by the [Human] Genome Project to actually be stillborn because people would be afraid of getting the information that otherwise would be of great advantage to them for medical purposes."[20]

> *We discard people who are too much trouble. These are entrenched values, eugenic values. We don't want to deal with people who do not fit the standard of physical attractiveness and normalcy.*
>
> Marsha Saxton, World Institute of Disability spokesperson, as quoted by Sally Lehrman in "Prenatal Testing Spurs Fear of Eugenics," GeneLetter 1(8), September 2000

As this book went to press, Collins told me that he was feeling cautiously optimistic about genetic discrimination legislation. Though the Senate has passed the bill, the House had still not taken it up. The issue, he says, is his number-one concern. "This is a matter of the greatest urgency. Already, people have lost health insurance and jobs as a result of finding out genetic information about themselves—information they needed for their own health care. Many other victims will be injured in the future if this is not dealt with. Genetic discrimination is unjust, and it could affect any of us."

After all, Collins adds, "We all have genetic glitches somewhere in our genomes."

Access

Even if people aren't genetically discriminated against by employers or insurance companies, another question remains. Will everyone be able to afford genetic testing? Some politicians and ethicists watching the issue worry that its cost may keep it from benefiting all citizens equally.

> *It simply isn't right that the very information which may lead to a healthier life and the prevention of a disease may also lead to the denial of health insurance or higher rates. Americans shouldn't have to make a choice between taking charge of their own care or keeping their insurance.*
>
> *U.S. Senator Olympia Snowe (R-Maine), 2003*

Consider this scenario. A treatment like preimplantation genetic diagnosis, for instance, costs as much as $4,000 over and above the required IVF procedure. Is that fair or right?

Troy Duster, a sociologist at New York University, and former head of the Human Genome Project's Ethics, Legal, and Social Implications (ELSI) board, says he sees uneven application of screening, fertility technologies, and other advances as inevitable. "We have a market economy, so there's your answer," Duster says. PGD will become increasingly available, and the people who can afford it and don't mind the hassle of the additional test will get it. It's better to face up to that fact now, he says, than fool ourselves into thinking that all technologies will be available to all the people across the board. That's never been the case.

I'LL TAKE ANOTHER BOY, PLEASE

The idea of selecting some embryos and disposing of others based on gender is anathema to most fertility specialists, who advocate the PGD procedure not for frivolous reasons, but for addressing serious hereditary diseases.

"Last I checked," says preimplantation genetic diagnosis pioneer Mark Hughes, "gender was not a disease."

A Virginia-based fertility clinic called the Genetics and IVF Institute (GIVF) is running ads in the Style section of *The New York Times* with quite a different message. It claims that people may want to remove boy embryos because of certain male-carried diseases, or sort out girls because of "family balancing issues." (That is a term GIVF founder Joseph Schulman apparently coined in an interview.) And GIVF says it has the technology—called MicroSort—to do it. According to company materials, MicroSort allows families to choose their "correct gender" baby in a couple of ways: Parents can use IVF to create multiple eggs and separate out the correct sexed embryos via PGD; or use sperm sorting and artificial insemination.

In the past, this technology has come under fire particularly from Asian activists, who say they know firsthand what gender selection does in a society. (In South-East Asian countries, demographers point to as many as 100 million "missing girls" killed by infanticide or neglect.)

Then again, as an article in the newsletter of the Center for Genetics and Society points out, it is clear that much of the sex selecting going on is by women who want girls.[21] Even in that case, ethicists worry. There is gender stereotyping, perhaps, in a parent who is investing money, time, and discomfort into producing a girl baby. One MicroSort customer was quoted as saying: "I wanted to have someone to play Barbies with and to go shopping with; I wanted the little girl with long hair and in pink and doing fingernails."[22]

Furthermore, in a study at Cleveland State University, researchers learned that more than 80 percent of women and 94 percent of men would use sex selection technologies to

{ make sure their first child was a boy. The second child, they
{ said, could be a girl. Given what psychologists know about
{ birth order—firstborn kids are more aggressive than oth-
{ ers—Cleveland study leader Roberta Steinbacher told a *New*
{ *York Times* reporter that she was concerned. "We'll be cre-
{ ating," she says, "a nation of little sisters."[23]

"I don't see any problem with parents giving a gene to their child
that will make their child resistant to diseases," Lee Silver, a professor of genetics at Princeton University, told an interviewer on
the radio series, *The DNA Files*. "I don't see any problem with parents choosing an embryo that doesn't have cystic fibrosis, or putting a gene into an embryo that protects that child from getting
AIDS or heart disease or diabetes or obesity. What I do see, though,
is the real ethical dilemma that's going to confront us in the
future—[which] is that this technology is very expensive. People
with money are going to be able to give their genetic enhancements
to their children, and people without money are not going to be
able to afford it."[24]

Silver has shared this concern in a variety of mediums. In an interview with the BBC that aired on January 1, 2000, he expounded:
"The problem here is that we're going into a world that is going to
be totally different [from] the world we live in today because there
really are no limits to what we can do genetically. Anything that I can
imagine in terms of changing genes in a baby, I could do. I could give
a baby the hearing ability of a dog or the eyesight of a hawk. I could
give that child anything—resistance to diseases of all different
kinds—and further in the future, as we understand more and more
about the human genome, we'll be able to increase that child's intellectual potential. All of a sudden you'll end up with a group of children who . . . are vastly superior at the genetic level to what I call the
naturals, those people who have not had genetic enhancements. I

don't think those people would be able to interact very well, and so they will stay apart from each other socially and ultimately they won't be able to breed with each other. That's exactly the way that new species get formed in nature. This, I think, is actually quite horrible. I think it's going to be a disaster because one group of people who is a different species to the other group of people will no longer have the desire or need to treat that second group of people with dignity and respect. And I think that's a pretty bad outcome although I don't see how we can stop it from happening."[25]

Says Mark Hughes: "All new technologies, from electronics to automotive to medical, cost money. It is unfortunate, but a reality of the American health system [is] that everyone is not treated equally. In European countries and Canada, if the disease is severe, the system covers it financially, just as it would an amnioicentesis or coronary bypass surgery. And, just like all technologies, the price falls as the usage increases and the methods improve."[26]

TWO GENES THAT COULD MAKE YOU FAT

Eating too much and burning off too little obviously puts on the poundage. But for the grossly obese, there may be a genetic explanation.

In late 2003, after studying the DNA belonging to 17,000 people in Iceland, deCODE genetics and its partner Merck & Co. announced they'd isolated two genes that may predispose people to obesity. They hope to eventually create drugs that act on the gene products created. Observers expect the effort to pay off within a decade.

"We have identified common [versions] of genes that contribute significantly to both of the principal processes involved in obesity—basic energy metabolism on the one hand and the regulation of appetite on the other," says Kari Stefansson, CEO of deCODE.[27]

Because of its unique tactic of gene hunting, deCODE has attracted quite a bit of attention. Several years ago, the company bought the rights to the health records and family history of Iceland, an isolated country whose population mostly descends from a small group of Vikings who settled there in the ninth and tenth centuries. Because of the island's relative isolation, the gene pool is particularly inviting to researchers searching for the gene variants behind disease and disorders.

With Merck, deCODE plans to release a series of genetic diagnostic kits for the genes it discovers through the Icelandic effort.

CHAPTER SIX

THE FOUNTAIN
OF AGING WELL

AS BUTCH CASSIDY said to the Sundance Kid, "Every day you get older. Now that's a law."

But, armed with all the growing knowledge of what makes us human, do we also automatically have to get cancer, heart disease, and become mentally decrepit? Is that a law?

Must we die at all?

It is true that, thanks largely to public health policies and antibiotics, human life expectancy has already seen improvement. For most of the time humans have been on this planet, most people were lucky to live past age 18. Now, that age is well over 80.

Life expectancy has increased by 2.4 years every decade for the last 160 years, without any help from geneticists at all. The question is whether we can push it further still.

Though there is wide debate over the claim, some scientists say a lifetime of 150 years old or even longer is accomplishable in our lifetimes. Scientist Steven Austad has put money down on it. He and biodemographer Jay Olshansky have bet that, by the year 2150, there will be people alive who have survived to an unprecedented 150 years of age.

"We picked the age of 150 precisely because we thought this was outside the possibility of achieving by incremental increases in medical care," Austad told me. "That age is only about 22 percent longer than the current record [of 122]."[1]

> *Though I am already seventy-five years old, I have reason to believe that I will personally benefit from DNA research.*
>
> James Watson, co-discoverer of the double helix, 2003

He and Olshansky put $150 apiece in an investment fund, with a plan to each add $5 to the pool every year. The bank will distribute the pool to the relatives of the winner in 2150—when it will be worth an estimated $500 million. For Austad's descendants to collect, the 150-year-old needs to be in fairly decent health, and needs impeccable proof of age.

Olshansky doesn't doubt that humans will make it to age 130 by then. Austad, meanwhile, points to recent success in increasing the lifespan of lab animals.

"Look at what Cynthia's doing," he says, referring to scientist-geneticist Cynthia Kenyon of the University of California at San Francisco. "Molecular geneticists like her are already helping identify the [proteins] that inhibit aging in [animals]."[2] Says Austad, "I can't believe we won't make improvements in human anti-aging treatments in the next 100 years."[3]

Worms Playing Tennis

Cynthia Kenyon also believes we can slow aging. And she has the worms to prove it.

"These worms aren't dead, they're moving around," she says. "They should be in the nursing home, but they're out playing tennis. It blows you away. They're like 450-year-olds who act and look like they're 60-year-olds. It just makes you wonder how far you can go."[4]

Until only recently, common scientific belief had it that there is nothing, aside from watching your diet and exercising, that you can do about aging. Scientists believed that like old rusty engines, humans just wear out. But Kenyon's experiments with roundworms (C. *elegans*) have riveted scientists in recent years.

She will likely go down in medical history as the researcher who has contributed most to the theory that aging can be delayed through gene manipulation. In essence, she has succeeded in doubling, then tripling, then sextupling the lifespan of her little rice-size roundworms, increasing their longevity far past their normal two weeks. Some of her worms live as long as twelve weeks.

Her team accomplished this feat by tinkering with three roundworm genes, called daf1, daf2, and daf16. This confirmed Kenyon's long-held suspicion that genes regulate aging, at least in simple organisms such as worms.

> *By design, the body should go on forever.*
>
> *Elliott Crooke, Stanford University*

"If you look at nature, what you see is really interesting: Different animals have remarkably different lifespans. Here is an example of three, small warm-blooded animals: a mouse, a canary, and a bat. You can see that a mouse lives about two years, canaries fifteen or so, and bats can live up to fifty years. How could they have such different lifespans? Well, they differ from one another, obviously, by their genes," says Kenyon."[5]

Just disabling the daf2 gene doubles the life of the little worm, Kenyon says. And it's not just extending the lifespan, it's extending "the good years." The elderly worms in Kenyon's lab don't look

FIGURE 6-1.
Cynthia Kenyon,
of the University
of California,
with her age-
defying round-
worms.

flabby and sluggish, like their unmodified peers; instead, they're shiny and squirming around like, well, whippersnappers.

What do the daf genes have to do with any of this? It turns out, says Kenyon, that they code for a hormone receptor that regulates insulin.

The gap between roundworms and humans is wide, of course, and converting Kenyon's research from worms to mice and then to humans will likely be a long and roundabout road.

> *You've got to have a mind of your own, that's for sure.*
> Centenarian Eva Fridell, quoted by Carol M. Ostrom in "How Did Eva
> Fridell Get to Be 110?" Seattle Times, November 18, 2003

Nonetheless, the work with the roundworms points to a tantalizing possibility. If aging is within genetic control in roundworms, it is conceivable that scientists will be able to come up with a drug or therapy that also interferes with human aging. It could, perhaps, reset the clock that triggers age-related demise. The question is when.

"People say they'll be surprised if this works in humans. I'll be surprised if it doesn't work," says Kenyon. "The only question is how long it will take to do it."[6]

She continues: "The premise is that we can slow down the aging process. And if we can do that, we can reduce the risks for all kinds of diseases. Cancer, heart disease, osteoporosis—the risks for all of these go up as you get old. But if we can slow down the aging process, we can reduce risk."[7]

Being age 90 and looking and feeling like 40, she says, is the ultimate goal. Kenyon and MIT scientist Lenny Guarente have cofounded a Cambridge, Massachusetts, company, Elixir Pharmaceuticals, to come up with anti-aging treatments that may take us on that road.

"We're not just talking about extending lifespan, we're talking about extending health span," she says.

Eos and Tithonus

With that, Kenyon hits on one of the main anxieties surrounding human longevity treatments. In extending human life, we must be careful to extend human health, too.

Maybe you remember from high school, or from reading *Bulfinch's Mythology*, the Greek myth of Eos and Tithonus. Eos, the goddess of the dawn, would fall hard for mortals from time to time, and she finally fell in love with the handsome Tithonus, prince of Troy. She begged Zeus to grant him immortality, and she got her wish. But she forgot to ask also for his youth.

> *Opinions go from nothing ever dies from old age, to every-thing dies from old age. We don't really know very well why people age to death.*
>
> Biogerontologist Steven Austad, "Staying Alive," Discover Magazine, November 2003

As the tale goes, Eos was mortified to see Tithonus grow older and older. When his hair turned white, she broke up with him, letting him roam the halls of her celestial palace alone. But when he lost the use of his limbs and could only babble at her, she turned him into a grasshopper.

Eighty and Loving It!

"Sixty Is the New Thirty" reads a cover of the AARP magazine, the one with the beautiful (and at the time of the photo, fifty-nine-year-old) Lauren Hutton on the cover. It is a breathless cover story and hardly scientific, but it makes you think.

Thirty has been a magic number. Aristotle once said "the human body is at its best between the ages of 30 and 35."

So if age 60 is the new 30, is 110 the new 80? Will it ever be?

"It remains to be seen," says Leonard Poon, director of the University of Georgia Gerontology Center, "if you pass the threshold of, say, 120, whether you could be healthy enough to have a good quality of life."

To point, Madame Jeanne Louise Calment, the world's record-holding oldest person, lived to 122. But in the end, her family members were literally propping her up for interviews.

Men grow old, pearls grow yellow. There's no cure for it.

Ancient Chinese proverb

Calment, however, lived independently well into her hundreds. Most centenarians, in fact, are even driving into their nineties and keeping up with lifelong hobbies. Is there a genetic reason for this? What do these old folks have that the rest of us may not? And is there a secret to prolonging youth, and not just years?

Old "youth" just seems to run in families, says Thomas Perls, a Boston University geriatrician who has been heading up the New England Centenarian Study, the largest DNA study ever of people age 100 and older. He is also a founder of Centagenetix, a Boston company that is hoping to find medicines to retard aging.

He points to a photo of Sarah Knauss who, before she died in 1999, held title to being the oldest woman in the United States. She is pictured, at age 119, alongside her 95-year-old daughter, her grandson, her great-granddaughter, her great-great-granddaughter, and her great-great-great grandson. In all, six generations of Knausses sat still for a single snapshot.

Pictures such as this will be increasingly common. The U.S. population now includes more than 40,000 souls age 100 or older. (In 1950, there were fewer than 2,300.) It is the country's fastest-growing demographic group.

> *I would say that our new approach to biology is going to lead the way to predictive and preventive and personalized medicating. That is a shift in medical practice that is going to radically transform the way medicine is practiced. Not only will it move us from being sick to worrying about how to stay well, but it will likely increase the lifespan of humans by ten or fifteen years.*
>
> **Biologist Leroy Hood**

ALL ABOUT THE TELOMERE

If you've been following anti-aging research at all, you've probably heard of a telomere. To refresh your memory, it is a little cap of DNA at the top of every chromosome, often compared to the little aglet on the end of a shoelace that keeps it from fraying.

James Watson, the co-discoverer of DNA, first spotted the telomere in 1972. After observing some DNA, he discovered that the chemical that helps DNA make copies—called polymerase—doesn't start at the top of the DNA strand every time, but several bases in. So the chromosome, presumably to keep valuable genes from being cut off during copying, has a string of words—TTAGGG over and over—at the top.

Every time the chromosome is copied, a little bit of the TTAGGG sequence is cut off, which is far better than a little piece of a real gene being deleted. Eventually, though, the chromosome runs out of telomere. According to one figure, you lose about thirty-two bases of telomere every year, which is why scientists say cells stop thriving at a certain point.[8]

But are people like cells? Do they have to age just because some of their cells do?

In 1984, scientists Elizabeth Blackburn and Carol Greider discovered a substance called telomerase. Its purpose apparently is to rebuild telomeres. Typically, the genes that code for telomerase are turned off in all your cells with exceptions such as the germ cells, the stem cells, the hair follicles, and other cells that keep dividing. But cancerous tumors know how to turn telomerase genes back on, which is why their cell line is essentially immortal.

So, do longer telomeres mean a decidedly longer life? Scientists aren't sure, although in 2003, Iowa State University researchers released a study that seems to support the theory. The storm petrel, a wild bird with a lifespan of up to thirty-five years, has significantly longer telomeres compared to shorter-lived birds.[9]

Could there be human populations that also have very long telomeres? It's "a very interesting possibility," says University of Utah geneticist Richard Cawthon. Such people, he said, might be a tough, especially long-lived group.[10]

"We are not trying to find the fountain of youth," Perls says. "If anything, we're trying to find the fountain of aging well."[11]

Most of the 750 participants in his centenarian study have aged well by any measure. "We have a small number of people, particularly guys, who do everything short of throwing an atomic bomb at their bodies and still live to 100," Perls says.

They eat lots of fat and sugar. They never exercised. Some have been smoking multiple packs a day for half a century. (France's Calment had smoked filterless cigarettes for more than a hundred years of her 122-year life.) They even seem to age more slowly. If you look at pictures taken throughout their lives, you will notice that centenarians generally look younger than their peers at every stage.

They just seem more vigorous. If you look at pictures of them growing up, they always look younger than their stated age. They just age more slowly.

Scientist Bard Geeseman, talking about many centenarians

Perls and fellow scientists say they are sure that these oldsters have genes that allow them to get away with things that would send most of us to early graves. But what are they, and where are they?

In all likelihood, there is a vast network of genes that helps people live to extreme old ages, he says. Some genes may slow aging throughout life.

Those genes have not been located, says Perls. However, a few age-related genes have turned up, according to his research. One lies smack in the middle of chromosome 4. Called the microsomal transfer protein gene, it appears to control how much cholesterol clogs up your veins.

If you have this gene—one of the "genetic booster rockets," Perls calls them—you are more likely to live longer. While it isn't a switch that goes on and off, the mere presence of this gene does seem to appear in centenarians more often than in the general population. The gene may have the effect of limiting, or at least delaying, the onset of such age-related diseases as Alzheimer's, stroke, heart disease, and cancer, Perls says.[12]

It remains to be seen, if you pass the threshold of, say, 120,
if you could be healthy enough to have a good quality of life.
Gerontologist Leonard Poon

Jumpstarting the Search

"When we finally are able to add significantly to our lifespans," says Cambridge University geneticist Aubrey de Grey, "we will look back and ask the moral question, why did we not do it sooner?"[13]

De Grey is perhaps the most outspoken biologist in the world. His Rasputin hairstyle and beard and his formal English demeanor belie his relative youth in this field—he is just forty years old. And he is an international rabble-rouser, a ruthless critic of the medical establishment's overly conservative approach to anti-aging.

De Grey has made headlines with his claims that, outside of the fringe, venture capitalists and pharmaceutical companies aren't investing enough in anti-aging research. He says that's primarily because there's no short-term profit in it.

"The funding isn't there," de Grey says. "But if we can do it in mice—significantly increase the years they are alive—this would be a result so impressive that it would trigger an immediate war on aging."

We now know all the processes that make up aging well
enough to target aging. And when you want to manipulate a
complicated system, you only have to understand it a limited
amount. You don't have to understand all of it.
Geneticist and gerontologist Aubrey de Grey

To this end, de Grey and colleagues have created the Methuselah Mouse Prize, named for the biblical figure claimed to have lived 969 years. In a prize potentially worth tens of thousands of dollars, scientists hoping to win must come up with the longest-lived laboratory mouse. (An alternate Methuselah prize will go to the late-intervention longest-lived mouse—that is, the mouse scientists waited until adulthood to treat.)

LIVING TO 500: A SEVEN-POINT PLAN

Aubrey de Grey at Cambridge University says the best way to extend human life is to fix things as they go wrong, not to try to slow the process of aging altogether. It's an engineer's approach, he says.

Here is the controversial seven-point plan that de Grey has been circulating among the world's biogerontologists. Learn how to fix these things, he says, and you will, in effect, be able to live forever.

1. Cell Therapy. Restore lost cells in tissues that tend to lose cells (such as the heart, as it ages, or the brain, in the case of Alzheimer's or Parkinson's disease).

2. Targeted Gene Therapy. Delete the genes in tumors that cause the lengthening of telomeres. Maintain the telomere elongation genes in rapidly renewing tissues that need it, such as the blood, skin, and gut.

3. Insertional Gene Therapy (Part 1). Replace the thirteen protein-coding mitochondrial genes with error-free ones so that mutations in the mitochondrial DNA (passed down the maternal line) can't hurt us.

4. Insertional Gene Therapy (Part 2). Add bacterial or fungal genes, such as those found in soil, to break down substances that tend to build up and limit lifespan. Such genes would code for proteins to break down cholesterol that clogs arteries—such as A2E, which causes macular degeneration—and various proteins that contribute to brain damage.

5. *Immune Therapy.* Destroy excessively aged cells. Some of these become actively carcinogenic. Just clean them out.

6. *Immune or Medicinal Therapy.* Break down the amyloid, the gunk that accumulates between cell walls as we age, particularly in the brains of Alzheimer's patients.

7. *Medicinal Therapy.* Break down the glucose links that form randomly between long-lived molecules that toughen skin and other tissues.

Getting a mouse to live to at least five years of age, instead of the normal two years, will be the first hurdle. Andrzej Bartke of the Southern Illinois School of Medicine managed to get his mouse, a genetically engineered critter named GHR-KO 11C (11C for short), to live 1,819 days, just short of five years. That is the equivalent of 150 to 180 human years. To accomplish this, Bartke engineered 11C with a gene that would limit the animal's production of insulin, leading to less age-related damage to the cells.[14]

> *At [age] 20 a man is a peacock, at 30 a lion, at 40 a camel, at 50 a serpent, at 60 a dog, at 70 an ape, and at 80 nothing.*
>
> **Baltasar Gracian**

Once imaginations are captured, de Grey says he is optimistic about extending human lives to age 120 or even 130, within decades.

"We now know all the processes that make up aging well enough to target aging," de Grey says. "And when you want to manipulate a complicated system, you only have to understand it a limited amount. You don't have to understand all of it."[15]

"If we manage to triple the life expectancy of a fifty-year-old," he says, "we are pretty much there." That is, typically a fifty-year-old could expect to live another thirty years. If we could triple his or her remaining years, then science would have ample time to catch up with even better, longer-lasting treatments.

The trick, he says, is going to be "repairing damage as it occurs."

It is sad to grow old, but nice to ripen.

Actress and activist Brigitte Bardot

An Engineer's Approach

Essentially, de Grey is advocating an "engineer's approach" to aging. Rather than trying to slow down the process of deterioration, you simply get better at fixing damage as it happens. This is the same as how you would keep an old house in good repair. You fix the roof when it has a leak; you paint the house when it needs it; you upgrade the wiring every few decades.

"This means that we should, in due course, be able to take people who are already middle-aged or more and rejuvenate them," says de Grey. "We will, in the first instance, only be able to do this imperfectly and incompletely, but that will be long enough to extend life span a bit. As time goes on, we will get progressively better at that.

"In fact, we will get better at an accelerating rate [as with all technology]. This means that eventually we will be getting better at fixing aging at a faster rate than time is passing. We will be encountering new things that go wrong with us at older ages, but we'll be fixing them faster than they arise." De Grey, to this end, has identified seven "strands," or areas of aging, that aging engineers might focus on in the future. They range from modifying genes to reduce the incidence of cancer to finding ways to replace cells that are lost to heart disease and Parkinson's.

"This might be difficult if it weren't for monkeys. They're fabulously similar to us and prone to age at least as twice as fast as us," de Grey says. "So we don't yet know what 200-year-old humans will die of, but we don't need to until we have some people that old—and by that time, we will for some time have had 100-year-old monkeys that we've been treating in just the same way that we treat ourselves: bad diets, no exercise, but all the life extension technology that we use on ourselves. And because those monkeys will have exhibited the same symptoms 200-year-old humans would, we'll have been working for a long time on fixing [those symptoms] in monkeys by the time humans get them. And when that occurs, we'll already know how to fix them well. By the time the first humans reach [age] 300, the same will be true by an even greater lead time.

> *How dull it is to pause, to make an end, / To rust unburnished, not to shine in use! / As though to breathe were life!*
>
> Alfred Lord Tennyson, in Ulysses

"This all [depends] on the monkeys getting the same problems that we get, but at under half the age, but that's a pretty safe assumption," de Grey says.

According to de Grey's vision, eventually scientists will reach a kind of "escape velocity," at which point anyone with access to the latest medical care could live almost indefinitely. "At that point we will die only from accidents, wars, homicide, et cetera."[16]

GETTING INTO THE FOUNTAIN OF YOUTH BUSINESS

Bruce Ames, of the University of California at Berkeley, is the namesake of the Ames test, a carcinogen screening test used throughout the world. But he is now known also as the founder of the company Juvenon, which is behind a dozen privately held anti-aging start-ups searching out the fountain of youth. Juvenon is studying the combination of acetyl-carnitine and lipoic acid—two supplements—in a clinical trial on heart disease.

Anti-aging researchers are flocking to such start-ups, and they are popping up all over the country.

BioMarker Pharmaceuticals in Campbell, California, is going the calorie-restriction route. Based on research that shows that severely restricting calories in mice lengthens their lives, BioMarker is trying to come up with drugs that duplicate a starvation diet's benefits without subjecting people to it. So is LifeGen Technologies, of Madison, Wisconsin. The firm is also looking into how calorie restriction affects gene expression, and the ways in which a drug might possibly mimic that.

Elixir Pharmaceuticals Inc., meanwhile, has partnered with large Hamburg, Germany-based Evotech, which has experience in medicinal chemistry. The goal is to come up with a drug, among others, that might act on the so-called IGF1 pathway, the human pathway most similar to the daf pathway affected in Cynthia Kenyon's mutant worms.

Slowing Aging Indefinitely

Richard Miller, a biogerontologist at the University of Michigan, has another perspective.

"Most kids, when they are growing up, go through a phase where the idea of getting old and dying is really scary. . . . I did, too. And most people grow out of it, and I didn't. . . . If you're interested in scientific mysteries, things that aren't yet solved, where people really need to use their intuition to discover what the important cracks are, aging is right up there at the top of the list as cancer biology was fifty years ago or infectious disease was 200 years ago."[17]

While he agrees with de Grey that most scientists are too pessimistic about longevity research and more funds are desperately needed, he doesn't think fixing things that go wrong may extend life by enough of a margin. Research shows, he says, that the average woman would only live to age 95 if cancer, stroke, heart disease, and diabetes were fixable. But if you could slow down her aging—as some scientists have done with mice by restricting their calories to the level of a near-starvation diet—she'd probably survive to the age of 115, and would be basically healthy up until the end.

FIGURE 6-2. Cambridge University geneticist Aubrey de Grey.

The idea that people can live longer as a result of a severely restricted diet is based on work done in 1935 at Cornell University. There, scientists discovered that calorie-restricted rats lived longer than rats fed regular diets. In the last sixty-five years, there have been hundreds of other studies showing similar results, and such organizations as the National Institute on Aging (NIA) spend several millions of dollars a year on related research. The calorie-restriction longevity technique may work because it lowers blood sugar levels, but scientists say they don't advise it for humans: It tends to make people miserable. However, a drug that duplicates the effect of lowering blood sugar is an oft-cited goal among anti-aging specialists.

"In general, I like the idea of fixing things," says Miller. "If someone has a broken arm, a cast is in order, and if someone has cancer, taking it out is a good idea. But in my view, none of this has much to do with aging research. So many things go wrong, more or less at the same time, in old individuals [such as two-year-old mice, ten-year-old dogs, and seventy-year-old people), that the notion that one can somehow fix all of them seems wrongheaded to me, particularly because at this point we don't have any really good ideas about how to stop any of the key problems—cancer, Alzheimer's, diabetes, heart attacks, hip fractures, and many other problems."

"The nice thing about anti-aging interventions," Miller continues, "is that, like caloric restriction and some genetic mutations, in unknown ways these postpone or retard nearly all of the adverse effects of aging at the same time. So I think it would be a good idea to learn more about how aging works to produce the diseases and disabilities of old age so that we could, potentially, figure out how to delay this process and stay alive and healthy for longer."[18]

Miller agrees with de Grey that one of the biggest problems facing aging research is a lack of funding.

> *Aging is, at its root, a side effect of being alive. More than 100,000 people die of it every day.*
>
> *Biogerontologist Aubrey de Grey*

"I think that if it were politically feasible to devote to aging research the same kind of funding that has gone into Alzheimer's disease research, or into AIDS, or into breast cancer, or 10 percent of the money that goes into the purchase of cosmetics—$45 billion per year in the United States—we would, in twenty to thirty years, have some pretty good ideas about how to delay aging in people," says Miller. "Actually testing these approaches in people would take a generation, though testing them in pets [dogs, for example] would be a good deal quicker."[19]

AN UNEXPECTED OBSTACLE FOR ANTI-AGING RESEARCH

As you've seen in this chapter, biogerontologists are all over the map in their theories for what exact treatments will work to delay aging and lengthen human life. But one of the biggest obstacles they face is the attitude people have toward the research. Biogerontologist Richard Miller calls it "gerontologiphobia."

"There is an irrational public disposition to regard research on late-life diseases as marvelous, but to regard research on aging, and thus on all late-life diseases together, as a public menace bound to produce a world filled with nonproductive, chronically disabled, unhappy senior citizens consuming more resources than they produce," Miller says. The same arguments were made 200 years ago, against penicillin, surgical anesthesia, and plumbing systems, he adds.

According to Aubrey de Grey, "[T]here's the plethora of arguments why curing aging might not be a good idea . . . but that's just a crutch to help people not get worked up about the perceived infeasibility. They'll be forgotten overnight when big progress is realized . . ."

Settling the Bet

Steven Austad is so sure he is going to win his bet with Jay Olshansky that he is talking about upping the ante. The bet—that come 2150, there'll be at least one person in the world who is 150 years old—will be worth about $500 million, with compound interest, when the time comes to settle it.

"My heirs or descendents will get all the money, or in the best case, I get all the money," says Austad. "I think this will be easy to win. We are making such rapid progress in understanding aging. We can make mice live six times longer than they normally do already. I'm convinced that within the next fifty years, there will be some serious [anti-aging] therapies available."

> *I have grandparents and parents who are getting up there in age. I can relate to this research and see myself still working in this field in twenty years.*
>
> Geneticist Cynthia Kenyon

How will it happen? Austad thinks that, of all the approaches scientists are currently taking, the one that biotech firm Elixir Pharmaceuticals, Inc. is taking is the most interesting. Elixir's approach involves modifying, with medication, perhaps, the IGF1 pathway in humans (equivalent to the daf gene pathway in Cynthia Kenyon's long-lived worms).

In twenty or thirty years, Austad guesses, scientists will be far along in animal trials. "We'll have discovered a lot more ways to make [lab animals] live longer. But living longer is not the same thing as aging more slowly," he says.

Then we'll be ready, he says, to figure out how to give the long-lived "the kind of life they want." That will require making sense of the "elaborate genetic symphony" going on in our body, where hundreds of genes related to aging need to produce their gene products, and everything must be precisely coordinated to work well.[20]

It's a tough job, but Austad predicts that at least one person will hit the 150-year mark by 2150. He's banking on it.

It's a Fact

Fact: Half of the American girls born this year, say some scientists, will live past age 100.[21]

Fact: On human chromosome 14, a gene called TEP1 codes for a protein that forms part of a chemical known as telomerase. Some cells turn immortal if you give them enough telomerase. That sounds like a good thing, but a cell line known as cancer also needs telomerase for its own immortality project.

Fact: Cancer cell lines, being immortal, are useful to scientists in the laboratory. Scientists cultivate and name them. One of the most prolific is the so-called HeLa cell line, derived from the tumor of a Baltimore woman, Henrietta Lacks, who died of cervical cancer in 1951. There are today so many HeLa cells in the world that they reportedly weigh more than 400 times what Lacks did when she was alive.[22]

Fact: The U.S. population now includes more than 40,000 souls age 100 or older. Compare that to 1950, when there were fewer than 2,300 centenarians alive. Today, the 100-plus crowd is the country's fastest-growing demographic group.

Fact: Animals with the fewest predators seem to survive the longest.

Fact: There is a gene on the middle of human chromosome 4, called the microsomal transfer protein gene, that codes for a protein that helps keep the arteries clear of clogs. If you have a particular variant of that gene, some scientists believe you stand a better chance of living longer.

Fact: While aging seems a fact of life, it isn't a fact of all life. Some bacteria are apparently immortal. Some large animals, such as the alligator and flounder, seem to be essentially immortal. They never reach "adult size." Rather, they just keep growing and never outwardly show signs of aging. The reason we don't see gators the size of Winnebagos prowling about is because they die from other causes, such as accidents.

CLOSING IN
ON CANCER

CALL HIM RICK. He was three when John F. Kennedy was shot, thirty when he got his first e-mail address, and forty-one the year the World Trade Center towers collapsed.

Born in 1960, he has seen some of the most crushing events in history—and some of its most important technological developments. And there is more to come.

In 2025, Rick will be age 65—and hitting senior citizenhood just in time to reap the benefits of genetics, genomics, bioinformatics, and nanotechnology. By then, some scientists predict, he'll be wearing a tiny gene chip under his skin that buzzes and glows at the first sign of cancer.

The tiny chip safely tucked under the skin a couple of millimeters next to his right elbow will perhaps be loaded with microscopic markers. These markers might chemically bind to DNA in his blood, DNA that will function as an early-warning sign for that cancer's particular known genetic mutations.

Perhaps it will glow one morning, and he will head directly to his doctor, who will analyze his blood to find out exactly what medication to prescribe. That medication—a pill, probably—will directly target his returning tumor, leaving the rest of his healthy cells intact.

Growth for the sake of growth is the ideology of the cancer cell.

Author Edward Abbey

Science fiction? Sure. But this vision isn't far-fetched.

"It's all to do with the coming together of digital telephone technology, bioinformatics, and genetics," Karol Sikora, the former head of the World Health Organization's Cancer program, told a journalist.[1] "The future is really about the little black box—this machine into which all the information about your genes and your behavior will be fed," Sikora said, explaining the idea. "Out will come a printout telling you what the correct therapy will be, and this is going to be different for each individual patient."[2]

Just as Leroy Hood's invention of the automatic gene sequencer enabled the mapping of the human genome, so will as-yet-unseen developments have an effect on cancer treatment.

Already, research into biochip technology, plus advances in target-specific drugs such as Herceptin and Gleevec, point to a future where, at the very least, many forms of cancer can be custom-treated and even managed.

A Golden Age for Cancer Research

In 1991, many cancer researchers were beginning to lose hope that humans would ever announce a victory in the "war against cancer" Richard Nixon announced with huge fanfare two decades before. Certainly, survival rates were no better than they were in 1971. And the more scientists learned about cancer, the more they realized it wasn't the simple but dreaded disease they expected.

Cancer, it turns out, is more than just cells that are unable to stop growing. Cancer is a collection of two dozen or more diseases, some hereditary, some caused by viruses or by mutations caused by chemicals, sunlight, or smoke.

But now, for the first time in decades, cancer researchers are optimistic. Most are hopeful that huge strides in fighting cancer are coming. Cancer research turns out to be the single biggest beneficiary of the mapping of the human genome.[3] Human Genome Project leader Francis Collins concurs with the idea. "I think that many of the earliest benefits will be reaped in the field of cancer, and hooray for that. It's already happening . . . with the use of microarrays to decide which patients with breast cancer need adjuvant chemotherapy, or with the introduction of Gleevec for the treatment of chronic myelogenous leukemia. Hard work by a legion of dedicated scientists has allowed the cataloging of the genes involved in cancer, making this field primed and ready for a genome approach to diagnosis, prevention, and cure. Other diseases—such as diabetes, heart disease, asthma, and mental illness—will ultimately benefit profoundly, too, but we have less of a foundation of knowledge to build on, so it will take longer."

Cancer is going to be very well understood in the next ten years. Certain cancers will be cured altogether.

Scientist David Galas

The National Cancer Institute predicts that by 2015 we will see "the elimination of death and suffering due to cancer."[4] "We'll start to see an impact, and by that I mean prolonging of life, in the next five years," Dr. Lee M. Ellis, an oncologist at Houston's M. D. Anderson Cancer Center, told *BusinessWeek* in 2003.[5]

Mike Stratton of the Cancer Genome Project, a U.K. effort that is searching for genes related to various common cancers, also agrees: "It would surprise me enormously if, in twenty years, the treatment of cancer had not been transformed," he says. "And when we look back we will see that those treatments emerged on the basis of the human genome sequence being announced today."

> *It would surprise me enormously if, in twenty years, the treatment of cancer had not been transformed. And when we look back, we will see that those treatments emerged on the basis of the human genome sequence...*
>
> Mike Stratton, Cancer Genome Project leader

Researcher David Galas of the Keck Graduate Institute, who made headlines by discovering the Alzheimer's and Werner syndrome genes a few years ago, told me we are well on our way to conquering this scourge. "Cancer is going to be very well understood in the next ten years, [and] certain cancers will be cured altogether. Maybe there are twenty or thirty types of cancer, and we're going to understand the fundamental mechanisms behind them. I'm sure of that. And we're going to find cures."[6]

WHAT IS CANCER?

Cancer happens when cells in some part of the body start multiplying out of control.

Unlike normal cells that grow, divide, and die in a more or less regimented way, cancer cells live longer than normal cells, rapidly form new abnormal cells, and even travel to other places in the body where they begin to replace normal tissue.

We now know for certain that all cancer develops because of some kind of damage to the DNA. Some damaged DNA is inherited. A mutation in BRCA1, which leads to a high susceptibility to breast cancer, is an example. Scientists have discovered other genes that are associated with cancer, specifically cancer of the colon, kidney, lymph node, pancreas, esophagus, rectum, and skin.

But frequently, cancer is not familial. Instead, it is spontaneous—a result of DNA damage that in turn resulted from exposure to smoke, alcohol, sunlight, some viruses, or toxins such as coal tar, asbestos, and hydrocarbons.

Such toxins either cause cancer by mutating a normal sequence or by affecting so-called oncogenes and tumor-suppressor genes. (To use a metaphor of a car, oncogenes promote cell growth. They are like a gas pedal. If mutated, the pedal's to the metal, and they send cells the signal to keep growing. Tumor-suppressor genes are the brakes. They restrict cell growth. If they are mutated, tumor cells keep growing without stopping.)

There are at least two dozen types of known familial and spontaneous cancers—some more aggressive than others. Not all respond to the same treatments. That's why different cancer patients often undergo very different treatments.

One in three Westerners develop cancer at some point in their lives, and one in five die as a result of it.[7] It is a battle with a lot of personal investment involved, and the big guns are out.

The American Cancer Society says scientists have learned more about cancer in the last ten years than they have during all the history of mankind. Whether that will translate into cures for cancer in our lifetime is the key question, and scientists hope the DNA sciences have the answer.

The Promise of Personalized Medications

Sarah Allen was 43 during the Christmas season of 2000 when her doctor told her that she had an especially aggressive form of breast cancer, and it was spreading.

Allen, a mother of four, rightly considered it a death sentence. She told a newspaper reporter that even as her heart sank, her oncologist looked hopeful. "There's this new drug . . . it's called Herceptin," he said. "We're going to give it to you as soon as you've had your surgery."

Herceptin is a new kind of drug, one especially created to target a genetic flaw that was contributing to her breast cancer. The flaw—too many copies of a gene called HER2—causes the overproduction of the HER2 protein on the surface of the tumor. (Only about 30 percent of breast cancer patients have this flaw, and their cancer is particularly aggressive.)

> *I attended a meeting at the National Cancer Institute and asked the question: "Have we found all of the oncogenes and tumor-suppressor genes in human beings, and is it time to stop looking and to start focusing on getting good drugs that make a difference?" Of the fifteen scientists in the room, some said we probably had found about 10 percent of the genes, while others said we probably had found almost all of them. The real answer is that we do not know. But the beginning of the answer is in the sequence of the human genome, which will tell us how to cure the cancer that begins in our own genes.*
>
> Scientist Arnold Levine, in The Genomic Revolution, (Washington D.C., Joseph Henry Press, 2002), p. 96.

Herceptin is a drug known as a monoclonal antibody, and it works by modifying a protein's production. Billed as a "biotic missile" by its maker Genentech, Inc., the drug manages to shrink cancer cells without killing healthy ones.

And several years later, Allen is still alive.[8]

Herceptin's way of working "is not like anything we have ever seen before" in cancer therapy, says Larry Norton of the Memorial Sloan-Kettering Cancer Center in New York City. Norton was one of the principal investigators for Herceptin. In his study, two women in the advanced stages of breast cancer were expected to live less than a year before taking Herceptin, he says. After taking the drug, they survived five and six years, respectively.

I wish I had the voice of Homer to sing of rectal carcinoma.

Scientist J.B.S. Haldane

"This is the biggest difference I have ever seen in advanced breast cancer," says Norton.

Herceptin is the first of a new trend of "targeted drugs." If you are diagnosed with breast cancer and tests show you have an over-abundance of HER2 genes, doctors know that you are an ideal candidate for Herceptin.

This fast-moving field will likely lead to personalized cancer treatments—special medical regimens precisely tailored to the characteristics of a person's tumor. For example, early studies on colon cancer patients have shown that a slight change in DNA can predict whether they'll suffer side effects from the powerful medicine Ironotecan. Other, similar studies are in the works that focus on lung cancer patients. Another study shows that it is possible to accurately predict how effective chemotherapy will be in many early-stage breast cancer treatments.

"In the past, we have not been able to reliably predict at the time of diagnosis which patients will experience a complete pathologic response to any chemotherapy regimen," says that study's lead investigator Lajos Pusztai, of the M. D. Anderson Cancer Center. "If our results are confirmed by larger ongoing studies, we soon may be able to select the best post-operative chemotherapy regimen for patients based on the gene expression profile of their tumors. This would maximize the chance of curing their disease, while sparing them from the toxic side effects of less effective treatments," he says.[9]

> *My veins are filled, once a week, with a Neapolitan carpet cleaner distilled from the Adriatic, and I am as bald as an egg. However, I still get around and am mean to cats.*
>
> Author John Cheever

The Poster Child for a New Kind of Drug

"There is, to my knowledge, nothing out there as exciting as Gleevec," Nobel Laureate David Baltimore told me. "It is the poster child for this new kind of drug."[10] A new kind of drug, he says, that treats cancer where it starts—genetically. As president of the California Institute of Technology (Caltech), he's in a good position to know.

Gleevec treats a certain kind of leukemia—called chronic myelogenous leukemia—with startling success. Trials showed it to be safe with few side effects. And nearly every patient who took it experienced a dramatic remission. A few patients developed a resistance to Gleevec, but most did not.

Because of Gleevec's outstanding performance, the Food and Drug Administration (FDA) rapidly approved it.

Gleevec's creator, Brian Druker, is also a poster child, but for a new kind of medical celebrity. The broad-shouldered, soft-spoken Druker has gotten press coverage worthy of Bill Gates or Steve Jobs. Druker is "the closest thing cancer research has to a hero," gushed *Wired* magazine.

The sequencing and analysis of the human genome, Druker says, was critical to the creation of Gleevec. "In the old days," Druker says, "by which I mean eight years ago, we knew about a handful of genes that might be involved in some cancer processes. Now we know about hundreds."[11]

FIGURE 7-1. Nobel laureate and Caltech president David Baltimore.

Gleevec works by changing the environment that cancerous white blood cells need to start multiplying uncontrollably. That environment is caused by a specific pair of genes fusing together, creating a protein that just keeps producing. Gleevec blocks that protein from overproduction.

Druker is quick not to label Gleevec (produced by Novartis) a "cure" for cancer, but "this truly does represent a new era in therapies for cancer.

We need cancer because, by the very fact of its incurability,
it makes all other diseases, however virulent, not cancer.

Author Gilbert Adair

"If you understand what drives the growth of a cancer you can target that abnormality specifically. In other words," he says, "you can disable the cancer without disabling the patient. You get there by a precise understanding of what drives the growth of a cancer. You identify the target, you develop a drug to inactivate the growth of that cancer, and then you end with a very special treatment.

"As we learn more and more about cancers," Druker says, "we're going to be able to develop drugs like this for each and every cancer."[12]

And that's the challenge now, says Caltech's Baltimore, for drug companies to come up "with other Gleevecs." Researchers are eagle-eyed, waiting for another such opportunity to develop.

Meanwhile, in Iceland . . .

Many of us have no idea who came before us three, or even two, generations ago. Not so for your typical Icelander. She can trace her family records back to the original Viking settlement of Iceland in the ninth century, thanks to the country's painstakingly recorded Book of Settlement.

That's handy information for Icelanders, sure. But for human beings, it's a treasure trove, too. Because Iceland is a remote, practically isolated island with a small population—about a quarter of a million people—its population is homogenous. The people are a lot more similar on Iceland than they are, say, in the melting pot that is California.

> *The older a cancer is, the worse it is. And the more it is involved with muscles, veins, and nutrifying arteries, the worse it is, and the more difficult to treat. For in such places incisions, cauteries, and sharp medications are to be feared.*
>
> Theodoric, Bishop of Cervia (1267 A.D.), in The Surgery of Theodoric Vol. 2, translated by E. Campbell and J. Colton (New York: Appleton-Century-Crofts, 1960), p. 26

No wonder Iceland is the place where myriad disease-related genes are being located. One of them, discovered in 1994, is called BRCA2, a gene that, when mutated, increases a patient's susceptibility to breast cancer by a factor of twenty. Scientists discovered BRCA2 after studying two Icelandic families, both of which can be traced to a single ancestor born in 1711.

Each family had a long history of frequent breast cancer, and cancer patients from both families shared the same exact mutation . . . a missing five letters (or bases) after the 999th letter of the gene.[13]

There is another mutation on the same gene, incidentally, that is common among Eastern European Jews (the so-called Ashkenazim). That mutation—a missing letter at the 6,174th position—is responsible for 8 percent of Jewish breast cancer cases alone. The Ashkenazim—because of their long cultural history prohibiting intermarriage with non-Jews—are another favored genetic hunting ground.

SHAPING UP FOR CANCER

Now that the human genome sequence is complete, researchers are hot on the trail of examining the proteins that the newly mapped genes code for.

One of those is a protein called Vinculin, and if you haven't heard of it yet, I bet you will.

At St. Jude Children's Research Hospital in Memphis, Tennessee, researchers discovered that Vinculin does something very interesting. It changes its three-dimensional shape so that a cell can move through its environment, rather than remaining fixed in one place. Researchers say Vinculin's ability to change its shape so that a cell with genes expressing it can move about reveals an important clue about how cancer cells are able to spread around the body.

"In other words, Vinculin is a critical protein that performs different roles in the body," says Philippe R.J. Bois, a St. Jude Department of Genetics fellow. "It is a master conductor of much of the cell's life, changing its shape to conduct the cell's business according to the cell's immediate needs." And Vinculin may actually help cancer cells spread.

Interestingly, Vinculin's shape-shifting abilities may also begin to explain how humans can produce so much complexity with only 30,000 or so genes.

"It was already known that cells can read certain genes in different ways to make different proteins," Bois said. "But these new findings significantly enhance our appreciation of the scope of protein function in the cell."

It is better not to apply any treatment in cases of . . . cancer; for, if treated, the patients die quickly; but if not treated, they hold out for a long time.

Hippocrates (460–377 B.C.)

Enter the Age of Predictive Medicine

The discovery of the breast cancer genes BRCA1 and BRCA2 in 1994 signified a turning point in breast cancer research. Both genes, when functional, are "tumor suppressor" genes. That is, they seem to control cell growth. But when they contain errors, out-of-control growth can result. Women who inherit faulty BRCA genes are at increased risk of getting breast and ovarian cancer—an estimated twenty times the risk.

By 1995, tests were on the market that could help people with breast or ovarian cancer in their family history determine if they, too, were at risk. (Myriad Genetics, Inc. in Seattle, makes and owns the exclusive rights to that test.) "The first big revolution that's starting to come now is very early diagnostics," says Leroy Hood, cofounder of Seattle's Institute of Systems Biology and the inventor of the automated sequencer.

"If you want to deal with cancer most effectively over the next five years, you want to call it cancer really early," he says. "You can't do preventive medicine without predictive medicine."[14]

There's an upside and a downside to such predictive tests, Hood says. Take the BRCA screening. For one thing, a positive result on the test doesn't mean a person will definitely develop breast or ovarian cancer, only that she is more likely to do so. Many who discover they have the mutation choose a bilateral mastectomy and an oophorectomy (removal of the ovaries) to decrease their chances of getting cancer. At least 15 percent of women who carry the mutation will never get the disease.

Testing negative for the mutation is misleading also. The majority of breast cancer cases—more than 90 percent—happen to women without either BRCA mutation. So a negative result is no guarantee that cancer won't strike later.

> *A cancer is not only a physical disease, it is a state of mind.*
> Michael Baden, New York City chief medical examiner, quoted by L. Johnston in "Artist's Death: A Last Statement in a Thesis on Self-Termination," The New York Times, June 17, 1979, p. 1

Finding out what causes the bulk of non-BRCA related cancers has been dogging researchers since the mid 1990s. But in late 2003, they finally got a break.

EMSY, the New Breast Cancer Gene

There are genes that code for proteins and genes that exist primarily to control other genes. In December 2003, researchers at the Cambridge University and Cancer Research U.K. found a control gene—called EMSY—that seems to be capable of shutting down healthy BRCA genes.

That begins to explain, perhaps, why many women without mutated BRCA1 or BRCA2 genes still get cancer.

"It's going to give us new lines of investigation and potentially exciting angles of attack," Cambridge researcher Tony Kouzarides told the BBC. "Discovering such an important new gene is very exciting and gives us the piece in the jigsaw we've been looking for. We'll now have a much more sophisticated image of the genetic changes triggering breast and ovarian cancer in people who haven't inherited a high risk of cancer, but develop it anyway."[15]

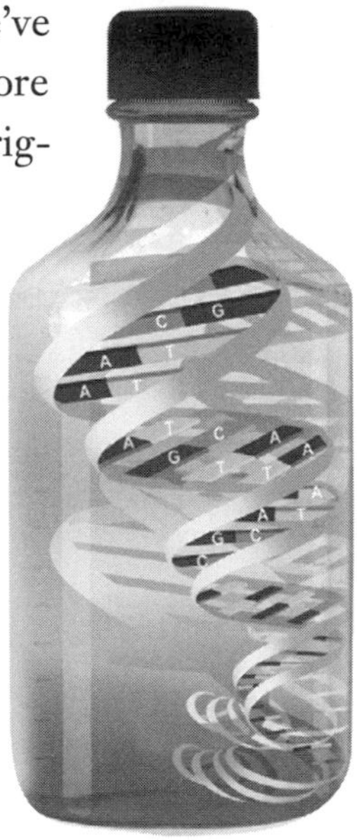

Kouzarides says he's reviewed hundreds of tumor samples, which showed that 14 percent of breast cancers—and 17 percent of ovarian cancers—contained extra copies of the gene. Yet, he didn't find the gene in normal tissue or in any other kind of tumor.

Says Professor Carlos Caldas, another researcher who worked on the study: "We've always thought factors that are important in inherited breast cancers should also be playing a role in other kinds [of cancer], and it's heart-

FIGURE 7-2. New kinds of drugs will treat cancer where it starts—genetically.

ening to know that we hadn't been barking up the wrong tree.

"It should help us in the development of new, targeted treatments against breast and ovarian cancer," says Caldas, "particularly as cancers with high levels of the new gene seem to behave in a similar way to inherited forms. EMSY could also form the basis for new types of predictive test."[16]

> *When the tumor extends its feet from all sides of its body into the veins, the sickness produces the picture of a crab.*
>
> Galen (130–200 A.D.), as cited in R. E. Siegel, Galen's System of Physiology and Medicine, p. 286

The presence of extra copies of EMSY seems to indicate a more aggressive type of cancer. Women whose tumors included extra EMSY copies survived only 6.4 years after diagnosis, compared to 14 years for women whose tumors had normal EMSY amounts. That means researchers will aim to come up with an EMSY diagnostic test soon, so doctors can predict how aggressive a cancer is likely to be and recommend the appropriately tough treatment.

Molecule of the Year

If you follow the health headlines, you've undoubtedly come across the gene named p53. The protein it codes for was even named "Molecule of the Year" by the journal *Science*. Scientists know that the protein created by p53 is key to cell growth. It's a tumor-suppressor gene that is supposed to stop cells from growing. When it's damaged, it can't do that. A mutated p53 generally means higher susceptibility to cancer. In fact, more than half of all human cancer cells contain a p53 mutation, which tells you how important and widespread it is.

Many researchers are continuing to look for ways to diagnose and treat cancer that's related to a p53 problem.

> *If you understand what drives the growth of a cancer you can target that abnormality specifically. In other words, you can disable the cancer without disabling the patient. You get there by a precise understanding of what drives the growth of a cancer. You identify the target, you develop a drug to inactivate the growth of that cancer, and then you end with a very special treatment. As we learn more and more about cancers, we're going to be able to develop drugs like this for each and every cancer.*
>
> *Gleevec inventor Brian Druker*

Another gene that researchers believe is associated with cancer is called ATM. Its general role is to control cell division. Though scientists do not yet understand why a mutated version of ATM can cause cancer, they do know that an altered form of this gene is linked to a childhood nervous disorder called ataxia-telangiectasia (AT). That disease makes children overly sensitive to radiation.

Another superstar gene you'll be reading about is p65. The mutated form of this gene seems to cause the overproduction of hormones related to breast and prostate cancer.

ROUNDWORM PROVIDES BREAST CANCER CLUES

If you followed DNA sequencing, you probably noticed that sequencing the roundworm and fruit fly almost seemed as big a deal as sequencing the human genome.

The reason, of course, is that the two organisms are such popular lab animals. Fruit flies and roundworms multiply rapidly and are easy to house, making them ideal for DNA experiments. And recently, the roundworm turned up yet again in important research—this time as the bearer of a gene that is remarkably similar to the BRCA1 gene that, when mutated, increases susceptibility to breast and ovarian cancer among women.

"It's nearly a decade since the BRCA genes were discovered and implicated in the development of breast and ovarian cancer, but we are still very much in the dark about how they function," says Simon Boulton, from the Cancer Research U.K. London Research Institute, which published the study.

The researchers discovered that in worms, the BRCA1 gene works in tandem with another gene, called BARD1. It's much the same way as the two genes interact in humans, he says.[17]

First, they switched off the two genes—simulating a mutation by causing the genes not to work properly. Then, they exposed the worms to cancer-causing radiation. When the worms got cancer, that proved that the two genes played a key role in DNA repair.

Professor Robert Souhami, director of clinical and external affairs at Cancer Research U.K., said: "Studying the BRCA1 counterpart in the worm will accelerate our understanding of how defects in this gene can lead to breast cancer and, in the future, will offer possibilities for prevention and treatment."

Researchers say that now that they have a model BRCA1 cell to experiment with in the worm, genetic analysis is much easier. "The detailed genetic analysis we can do in cells of the worm is not really possible in more complex human cells," Souhami says.

Cutting Off the Blood Supply

One of the hottest buzzwords in biotechnology is angiogenesis. Essentially, it means "artery building." Tumors have to do a lot of artery building to feed themselves.

As Galen (130 200 A.D.) noticed when he named it, cancer ("carcinoma" means *crab* in Greek) is crab like it spreads by building and sending out new arteries to feed its ever-growing need for more blood. The three-decades-old theory behind anti-angiogenesis treatments is that if you cut off the blood supply, you stop the tumor.

> *The whole angiogenesis strategy holds the promise of providing the cures for a number of cancers.*
>
> *MIT cancer researcher Robert Weinberg*

Today, more than 1,000 laboratories and 300 biopharmaceutical companies are studying anti-angiogenesis. There have been more failures than successes. But one of those companies, Genentech, announced what many consider to be a significant breakthrough in 2003. In human phase III trials, Genentech's Avastin drug starved tumors and prolonged colon cancer patients' life by more than 30 percent.

The results validated the research of Judah Folkman, who in 1971 first hypothesized that tumors depend on the growth of new blood vessels. Though few took his work seriously at the time, many experts now consider him to be a medical pioneer of the highest order.

Avastin works by targeting a protein that cancer cells make, called VEGF (vascular endothelial growth factor). By neutralizing that protein, Avastin slows cancer cells' ability to recruit new vessels.

"This is a major advance in the development of cancer therapy," says Dr. William Li, medical director of the Angiogenesis Foundation. "Avastin is the first angiogenesis inhibitor drug to show, in large-scale, well-designed clinical trials, that stopping the tumor blood supply is a valid approach to treating cancer. This marks a milestone for improving cancer treatment."[18]

"The whole angiogenesis strategy holds the promise of providing the cures for a number of cancers," says MIT's Robert Weinberg, founding member of The Whitehead Institute for Biomedical Research and one of the nation's foremost cancer authorities.[19] There are more than sixty other anti-angiogenic drugs in human testing, and two drugs that are right on the heels of Avastin in the approval process. One is AEterna Laboratories' Neovastat, targeting kidney cancer and non-small cell lung cancer. Another, called PTK787, from drug maker Novartis, is in testing for colorectal cancer.

> *We are slowly but surely having an impact on selected cancers. We just have to be patient.*
>
> *David Baltimore, Caltech president*

Thalidomide is another surprising anti-angiogenesis candidate. Vilified in the 1950s because it was shown to cause birth defects when prescribed to pregnant woman, thalidomide seems to also inhibit vessel growth. New Hampshire toddler Melanie McDaniels was among the first to realize the newfound benefits of thalidomide. After two surgeries, her brain tumor kept growing. But then her doctor enrolled her in a thalidomide trial, requiring blood infusions for a week every twenty-one days.

"We decided that, if anything, it might carry her along till they found something that would work," her father, Paul McDaniels, told a correspondent for *BusinessWeek*. But the thalidomide worked. And the tumor stopped growing.[20]

HAVE CANCER? TAKE AN ASPIRIN

You've heard of aspirin as a preventive treatment for heart disease, but as a treatment for cancer?

The idea isn't as far-fetched as it sounds. Researchers at the University of Connecticut working with mice recently discovered that the rodents with fastest-growing and fastest-spreading types of breast cancers produced larger amounts of a protein called COX-2. It is the same protein that common painkillers target.

COX-2's main product is an enzyme called PGE2, a potent chemical that helps tumors build the blood vessels they need to survive.

It turns out that several commonplace household drugs have been shown to inhibit COX-2 production, and two of them are aspirin and ibuprofen. At least in theory, they could help patients prevent or slow down the growth of breast tumors, though researchers would want to create a more specific drug without the side effects caused by the long-term use of painkillers.

"It's pretty awesome," says Timothy Hla, director of the University of Connecticut's Center for Vascular Biology and the study's senior author. "It's hard to believe something that simple might help fight cancer."

Many more tests are needed, he said, to see whether medicines that inhibit PGE2 can combat tumors in humans.

While anti-angiogenesis drugs show much promise with some patients, there are challenges ahead. Such drugs will likely have to be prescribed in combination, as they block only one or a few vessel-growth proteins at a time. There are at least a couple of dozen proteins involved in blood vessel growth, and many more clinical trials will be required to find out which drugs work best in concert or alone.

That's the exciting phase we're in. It's like a Polaroid. We're beginning to see what's possible.

Cancer map researcher Todd Golub

Nonetheless, anti-angiogenesis drugs are an exciting development, and what makes them possible is our increasing understanding of our genes and the proteins they make.

THE BENEDICT ARNOLD OF CANCER

Everyone has genes that help their bodies heal wounds. But if researchers at Stanford University are correct, a cancerous tumor could turn the same genes against you.

"This is a feature we can find early on in the disease and it could change the way cancer is treated," says Howard Chang, MD, PhD, the lead author of the study.[21] Researchers discovered the role wound-healing genes play in cancer by looking at cancer from a new perspective. Instead of dissecting a tumor and testing it to see which genes are most active—a technique that has allowed researchers to identify legions of genes involved in cancer—they examined what genes were involved in wound healing and checked to see if they were the same ones that were active (i.e., working and producing proteins) during cancer growth.

"Wound healing is a process that allows cells to break normal constraints on their growth and cross boundaries. If a cell can access that program, that's a good environment for cancer," Chang says.

And just like cancer, wound-healing genes rely on angiogenesis processes to bring a new supply of blood to a region.

To sum up, some tumors apparently use the body's natural wound-healing capabilities to help cells grow faster and move about. Same goes for blood vessel building, or angiogenesis. Growing new blood vessels is important in healing, but tumors can't grow without it.

The hypothesis turned out to be correct in certain kinds of cancer, specifically in liver cancer.

The effort to determine, very specifically, the environments in which certain cancers thrive is the first step in coming up with extremely targeted cancer drugs. Herceptin, for breast cancer, is an example of a powerful one. The next step: figuring out how to treat tumors that produce these wound-healing proteins. Chang says that scientists have a strong understanding of wound healing, so perhaps they will be able to throw a wrench in the process to stop a cancer from spreading.

"There are drugs coming out that block blood vessel growth, so perhaps those drugs should be targeted to this population of patients," Chang said.

A Global Cancer Map

Without computer technology to automate it, the sequencing of the human genome wouldn't have been completed in our lifetime. And technology continues to play a major role in virtually all the DNA sciences.

Nowhere is this clearer than in the global cancer project designed by MIT researcher Todd Golub. Essentially, Golub is comparing the genes in thousands of tumor samples in order to classify them. This is critical because cancer cells that look similar under the microscope and cause similar symptoms could in fact be radically different. And that in turn means that different drugs are necessary to treat them.

To do this, Golub is relying on microarray technology, the so-called DNA chip. Developed at Stanford in the early 1990s, a DNA chip is a postage-stamp-size glass wafer with a grid of up to 16,000 spots to hold DNA. In each spot, each piece of DNA acts as a probe. Essentially, as a *Wired* magazine article described it, you can look at the tumor's genes and tell exactly who is misbehaving.[22]

Here's how it works: Recall that any given stretch of DNA will always bind with its DNA or RNA match. TAG, for instance, will always stick to ATC (if DNA is used) and AUC (if RNA). So, to use a DNA chip, scientists merely liquefy a tumor, pour it over the chip, bake it at 113 degrees for a few hours, and presto! The products of the tumor genes—the proteins the genes code for—automatically attach to the chip. And they attach in amounts that proportionally reflect how active the genes are.

The more active certain genes are—that is, the more protein they produce—the more active they presumably are in the tumor in question.

Some scientists are skeptical that this is little more than a mental exercise—if there are too many subcategories, the data doesn't mean much. Tailoring drugs too specifically to every possible subvariety of cancer may be too expensive for any drug company to manage. Then again, that might be the only way to treat cancer—narrowly. And perhaps the categories won't be as tiny as some fear. Golub doesn't think so. "There will be some rules that cut across all types of cancers," he says. One DNA chip experiment showed that a wide range of tumors shared as many as seventeen (active) genes.

"It's a result that popped straight out of the microarray statistics." Adds Golub: "That's the exciting phase we're in," he says. "It's like a Polaroid. We're beginning to see what's possible."[23]

It's a Fact

Fact: Cancer has been tormenting man throughout recorded history. The earliest written reference to it dates back 5,000 years. A collection of seven ancient Egyptian papyri detail various treatments for cancer, from surgery and pills to magical incantations and prayers.

Fact: Why did researchers call the new gene EMSY? Professor Kouzarides's team, looking for DNA sequences that interact with BRCA2, discovered that one sequence includes these amino acids: serine, isoleucine, serine, threonine, glutamic acid, and arginine. In scientific parlance, the abbreviation for these spells S-I-S-T-E-R. So the gene is named after a cancer nurse named Emma, the sister of Dr. Luke Hughes-Davis, who discovered the gene.

Fact: For the first time in decades, cancer researchers are optimistic, believing cancer research is at a turning point. Cancer research turns out to be the single biggest beneficiary of the mapping of the human genome.

CHAPTER EIGHT

CLONING AND STEM CELLS

EVEN IF YOU don't know much about cloning and the heated viewpoints surrounding it, it's a sure bet you've heard about Dolly.

That sheep made headlines back in 1997 when embryologist Ian Wilmut and colleagues at Scotland's Roslin Institute revealed she was the first mammal cloned from a cell of an adult animal.

The first mammalian clone. It was such a big deal that many scientists unfamiliar with the project didn't even believe it at first. Since then, scientists have managed to clone many other kinds of mammals—a horse, a cow, a deer, a cat, a monkey—essentially creating a new animal that is virtually the identical twin of a single "donor" adult animal. It isn't even news anymore. Instead, it's taken a backseat to the real questions now on everyone's mind: When will someone clone a human? And what will it mean when it happens?

As this book was going to press, a U.S.-based fertility specialist named Panayiotis Zavos had just announced to the world that not only had he cloned a human embryo, but he had implanted that embryo in a woman's uterus. By the time you read this, the world will know whether the woman actually became pregnant and, if so, the welfare of that baby at birth.

But even before they knew whether Zavos's claim was for real, the announcement outraged the scientific community. Lord May of Oxford, the president of Britain's Royal Society, told Reuters that "the advocates of the reproductive cloning of people seem more motivated by the publicity of carrying out such experiments, in the face of overwhelming scientific and medical opinion, than by a genuine regard for the plight of the human guinea pigs that would take part."[1]

All of the data on animal cloning demonstrates exceptionally high rates of fetal loss, abortion, and neonatal deaths, and many cloned animals have devastating birth defects.

Obstetrics scientist Gerald Schatten

Three Different Kinds of Cloning

The term *cloning* means different things to different people.

Scientists use it as a blanket term for any process that can make a copy of genetic material, regardless of whether it's a fragment of DNA or an animal.

Cloning sometimes refers to the process of copying a fragment of DNA so that there are enough identical fragments for a scientist to study. For instance, criminologists use a method of DNA cloning called polymerase chain reaction (PCR) when they need to make many copies of a tiny bit of DNA found in blood, hair, skin, or semen at the scene of a crime. That is a widely accepted form of cloning used in labs around the world every day.

Then there is reproductive cloning, a technology for creating an entirely new animal (the clone) from the genetic material of an existing animal (the donor). Dolly was created with reproductive cloning technology.

HOW TO MAKE A WHOLE NEW ANIMAL

Dolly and most other cloned animals are created through the reproductive cloning process, in a procedure called "somatic cell nuclear transfer." Essentially, a scientist uses a tiny needle to pull DNA material from the nucleus of a donor cell and transfer it into a hollow egg. That egg has had its own nucleus and genetic material removed. To get the cell to start dividing, lab technicians then "stimulate" the egg with a chemical bath or a jolt of electricity. Once the egg has gone through several divisions successfully, scientists transfer it to the uterus of a female animal, who carries it until she gives birth to it.

Dolly and other clones are often characterized as the "twin" of the animal that donated the genetic material, but that is not true. The DNA in the nucleus of the clone animal might come exclusively from the donor, but recall that every egg carries some DNA outside its nucleus—a few genes of so-called mitochondrial DNA. So the clone also has some genetic material from the egg part of the equation.

Not incidentally, that genetic material may be mutated or damaged in the laboratory process, which could explain why many cloned animals don't survive to a healthy old age. Finding out exactly why cloned animals have so many problems is an enormous challenge that must be addressed, particularly if society considers cloning animals for food or using cloning technology to make medicine or babies.

> "There are a lot of questions to ask about cloned cells before
> you can justify putting them in a patient," says Ian Wilmut,
> who led the effort to clone Dolly.[2]

Therapeutic cloning—the cloning of human embryos to harvest stem cells for medical uses—is accomplished the same way as reproductive cloning. The difference is, instead of implanting the embryo in a woman, scientists destroy it so that researchers can extract its stem cells, which are master cells capable of morphing into different kinds of cells, such as those in the brain, muscles, or other organs, and which might be used for medical treatment.

The Dangers of Cloning

After Dolly appeared in 1997, most observers thought it would be just a matter of time before some renegade scientist somewhere in the world created a human clone in the laboratory.

Though that is true, coming up with a clone, human or otherwise, is an exceptionally difficult and inefficient process. Dolly, for instance, was the only living result out of 277 tries. Like most animal clones, she was created through a technique called somatic nuclear transplant. Essentially, scientists hollow out an egg, fill it in with genetic material from a cell from a donor, then fuse the two together to get it to multiply. After a certain number of cell divisions, scientists implant the fertilized egg into a female that will carry the fetus to term. But it's a hit-or-miss procedure. Of the twenty-nine cloned embryos that resulted from the sheep experiment, Dolly was the only one to be born.

> ### GOOD-BYE, DOLLY
> Dolly's birth snagged front-page headlines, but many of us
> missed the news of her death. Dolly died by lethal injection
> on Feburary 14, 2003, after suffering lung cancer and crip-
> pling bouts of arthritis.

Six years is about half the normal life span for a sheep of her type. Some scientists suspect that there is something inherent in the reproductive cloning process that leads to sick and/or oversize animal clones, and they hope to find the explanation.

In her brief lifetime, Dolly gave birth to six lambs the normal way.

And cloning can be hazardous to a little clone's health.

In *USA Today*, Gerald Schatten, vice chairman of obstetrics at the University of Pittsburgh School of Medicine, gave a warning: "All of the data on animal cloning demonstrates exceptionally high rates of fetal loss, abortion, and neonatal deaths, and many cloned animals have devastating birth defects."[3]

Clones that are born typically are oversized, and they often suffer from arthritis, cancer, and other diseases. This is a big reason why the majority of scientists take a stand against so-called human reproductive cloning, the type of cloning used to create a new animal from the genetic material of an existing one. It is simply too dangerous.

According to research from the Whitehead Institute for Biomedical Research at Massachusetts Institute of Technology (MIT), even when clones appear normal, they may still have genetic disruptions that can cause unpredictable medical problems. "Disruption of those genes in humans could cause [conditions such as] mental retardation," says Kevin Eggan, one of the researchers on the Whitehead team.[4]

CLONING—A TIMELINE

1952 Scientists create the first cloned animal, a tadpole.

1972 Scientists clone the first gene, using a yeast that incorporates the gene into its cells and multiplies.

1976 The first mice containing human DNA are born.

Scientists at the Salk Institute in La Jolla, California create these "transgenic mice" so that they can more accurately test human medicines on their lab animals.

1978   "Test tube baby" Louise is born. She is the first child conceived through in vitro fertilization. That is, the sperm and egg responsible for creating the fetus met in a test tube, not a woman. Today, more than a million test tube babies have been born in Western countries.

1996   Dolly is conceived. A year later, in 1997, the world's first cloned mammal is revealed to the world by Roslin Institute scientists. She is euthanized in 2003 after suffering cancer and arthritis.

1998   University of Hawaii scientists clone more than fifty mice from adult cells. Japanese researchers create eight cloned calves.

2001   Britain becomes the first country in the world to legalize the creation of human embryos—not to create living human clones, but to create embryos whose "stem cells" can be taken for experimental use. (Under new regulations, the clones must be destroyed after fourteen days, and it is illegal to create live babies by cloning.)

2002   At Texas A&M, scientists clone a calico-and-white cat. She is named "cc," short for copycat.

2003   The American Medical Association (AMA) endorses cloning for research, but says doctors who ethically oppose the procedure may refuse to perform it.

Ian Wilmut, the scientist who cloned Dolly and now a leading commentator on the controversy, agrees that cloning for reproduction, while potentially worthwhile for the animal husbandry business, is just too dangerous to try with humans. Says Wilmut, "It surely adds yet more evidence that there should be a universal moratorium against copying people. How can anybody take the risk of cloning a baby when the outcome is unpredictable?"[5]

> *I have been dealing in reproductive medicine for the last twenty-five years and never failed. I do not intend to fail now.*
>
> Fertility doctor Panayiotis Zavos, speaking about his claims to have implanted a cloned embryo into a woman (from "Cloning Doctor Brushes Off Criticism," Reuters News Service, January 23, 2004)

That probably won't stop a renegade scientist, such as Panayiotis Zavos, from trying. And if he doesn't succeed, someone else will likely try, observers say. "It is absolutely inevitable that groups are going to try to clone a human being," says Thomas Murray, a bioethicist and president of the Hastings Center, a bioethics think tank in Garrison, New York. "But they are going to create a lot of dead and dying babies along the way."[6]

But many of the same scientists are nevertheless quick to urge that the ban not be extended to therapeutic cloning and the harvesting of stem cells to cure disease. Scientists say the use of stem cells to grow new cells has the potential to treat or cure dozens of degenerative diseases, from heart disease to Parkinson's to kidney failure.

STEM CELL TREATMENTS FOR PARKINSON'S DISEASE

Could stem cell technology help reverse the physical decline suffered by Parkinson's patients?

Many scientists say yes. Renowned stem cell expert, Swedish researcher Olle Lindvall, says he expects to be able to transform stem cells into the dopamine-producing neurons Parkinson's patients so badly need. But it will take time.

"Stem cells could be potentially useful for the treatment of Parkinson's disease—but it's a very difficult problem to generate large numbers of dopamine-producing neurons, which are the cells we need," Lindvall says. "I am convinced that stem cell technology can become in the future a cure for conditions leading to brain injury—but I think we have a long way to go."[7]

Scientists have had some success treating Parkinson's in animals using stem cells from aborted animal fetuses, Lindvall says, but those stem cells aren't as effective as ones harvested from very early embryos of just a few days old.

And there is another possibility. Lindvall's research has shown that the brain of a rat, after a stroke, actually produces new brain cells that travel to the damaged area. Perhaps that process—plus some encouragement from drugs and combined with stem cells treatments—may be the eventual treatment for Parkinson's.

"I am convinced that therapeutic cloning offers health opportunities that you could not attain in any other way," says Wilmut, adding that it shouldn't be banned along with reproductive cloning.[8]

Responding to fears that if therapeutic cloning is allowed, some renegade researcher may decide to implant a cloned embryo, rather than destroy it, Wilmut says, "We can't stop this valuable research from going forward for fear of the few bad apples out there. That's why there are laws."

Cloning for Stem Cells

"Our intent is to use this technology to generate stem cells to treat serious and life-threatening diseases, not to create a child," says Robert Lanza of Advanced Cell Technology (ACT). He told me that his is one of the very few efforts in the world that has successfully cloned a human embryo. ACT is one of the very few private companies in the United States that kept working on stem cell research after the U.S. government dried up federal funds for the procedure.[9]

Think of an embryonic stem cell as a kind of master cell, an early-stage cell that retains the ability to form almost any kind of cell or tissue type in the human body. With a little chemical encouragement, a stem cell can turn into new heart muscle for heart attack victims; new neurons for stroke, paralysis, or Parkinson's patients; or new insulin-secreting pancreas cells for diabetics. Down the road, scientists believe it will be possible to create such complicated structures as blood vessels, liver tissue, and whole kidneys. In fact, ACT scientists have already succeeded in building tiny cow kidneys that could be used for kidney transplants. It isn't hard to envision, Lanza says, a future where pretty much any kind of organ or tissue could be engineered to replace those damaged by age, injury, or disease.

RESEARCHERS CREATE JOINT FROM STEM CELLS

Scientists say they've managed to build the ball-structure of a joint from adult stem cells retrieved from a rat's bone marrow.

Working at the University of Illinois in Chicago (UIC), researcher Jeremy Mao says he succeeded in transforming stem cells into the bone and cartilage tissue of a human jaw joint. "This represents the first time a human-shaped [jaw joint] with both cartilage and bone-like tissues was grown from a single population of adult stem cells." Mao, who is director of the university's tissue engineering laboratory and a professor of bioengineering and orthodontics, was speaking at a UIC press conference on December 1, 2003.

"Our ultimate goal," adds Mao, "is to create a [jaw joint] that is biologically viable—a living tissue construct that integrates with existing bone and functions like the natural joint." So far tested only in animals, the procedure promises to lead to technology that may help doctors replace hip, knee, and shoulder joints that are damaged by arthritis or other disorders.

The procedure is relatively straightforward. First, researchers prodded the stem cells, with the appropriate chemicals, nutrients, and growth hormones, to turn into cells capable of producing cartilage and bone. Then, they separated the cells into two layers and poured them into a mold created from the jawbone of a human cadaver. After a few days, researchers were delighted to discover that they had what they were looking for—joint-shaped tissue that had bone on the inside and cartilage on the outside, just like a human joint.

Tests confirmed that the engineered tissue actually was bone and cartilage, with all the typical components they have, including calcium.

Generally, adult stem cells—that is, stem cells found in bone cartilage—aren't as versatile as stem cells harvested from embryonic tissue. But this study suggests adult stem cells may be more useful than previously thought.

"It's not science fiction at all. This field is moving ahead so phenomenally quickly that by the time the baby boomers age, this could be routine stuff," says Lanza, adding that scientists have already developed techniques that could cure macular degeneration, a mal-function of the retina that leads to poor vision and blindness in more than 1.7 million Americans. But getting such techniques into the clinical trial stage is quite another matter. "We only have eight sci-entists and thanks [to the federal funding ban], there are times when we can barely make payroll. My hope is that as soon as we can show that we can cure diabetes in dogs, people will clamor for this. And then," Lanza says, "everything will change."[10]

The Debate

In the meantime, the debate rages on. President George W. Bush has made no secret of where he stands on the issue of stem cell research. "We recoil at the idea of growing human beings for spare body parts, or creating life for our convenience," he said in his August 2001 televised address to the nation.

And the United Nations, which was prepared to enter into a long-term treaty to stop scientists from pursuing human reproduc-tive cloning, instead hit a deadlock when the United States, the Vatican, and fifty Catholic countries pressured the U.N. to ban ther-apeutic cloning, too. The whole issue is now shelved until delega-tions have put more study into it. The treaty won't come up for discussion again until 2005.

> ### TOP TEN CAUSES OF DEATH IN THE UNITED STATES
>
> The top killers of Americans, according to the Centers for Disease Control and Prevention, are as follows:
>
> *1.* Heart Disease
>
> *2.* Cancer
>
> *3.* Stroke

4. Chronic lower respiratory diseases

5. Accidents

6. Diabetes

7. Pneumonia/flu

8. Alzheimer's disease

9. Kidney disease

10. Suicide

Deaths: Final data for 2001 (National Center for Health Statistics/Centers for Disease Control and Prevention).

The reaction among scientists varied widely. Some, like Bob Ward, spokesperson for the Royal Society in the U.K, said, "No decision is better than the wrong decision."

In other news reports, some scientists say they felt cheated. "Rather than ban the thing we all agree on, we end up with no ban, because the extremists refuse to compromise," says Larry Goldstein, a stem cell researcher at the University of California at San Diego.[11]

> *It would be dangerous and scientifically irresponsible. I don't know of a reputable scientist who'd consider using this technology to clone for reproductive purposes.*
>
> Therapeutic cloning scientist Robert Lanza, quoted by W. Goldman Rohm in "Seven Days of Creation," Wired, January 2004

Some observers are concerned that the U.N.'s delay in banning human reproductive cloning gives scientists hoping to make a big name for themselves or a fast buck from creating human clones some sort of safe haven.

Parthenogenesis: An Easy Answer?

In January 2004, Lanza and his fellow scientists at ACT made an announcement: They had succeeded in bringing a human embryo to the point of 100 cells through a technique called parthenogenesis. This was important news.

The same kind of reproduction that occurs in snakes and some birds, parthenogenesis leads to the creation of embryos (or "parthenotes") that don't include the male chromosomes required to make a placenta, so they likely could never be born as a living human. Perhaps stem cells created through this method won't be as controversial, and it could become the primary way stem cells are harvested for therapeutic purposes.

> *I strongly oppose human cloning, as do most Americans. We recoil at the idea of growing human beings for spare body parts, or creating life for our convenience. And while we must devote enormous energy to conquering disease, it is equally important that we pay attention to the moral concerns raised by the new frontier of human embryo stem cell research. Even the most noble ends do not justify any means.*
>
> *U.S. President George W. Bush, August 9, 2001*

"This is an ongoing research project and there are many steps ahead, including developing the cells into viable therapies," says Lanza.

The whole issue of a looming United States and, possibly, United Nations ban on cloning riles Lanza, who claims that stem cell therapy is the best shot that millions of Americans have to adequately treat their degenerative disorders. "It's unconscionable," says Lanza, "for Catholics and other evangelists to deny others the right to receive medical therapies. It's the whole issue of church and state. Who is the government to be taking sides in these religious debates? They should be looking out for the health and well-being of their citizens."

> *We have patients dying for lack of transplantable tissue on one side of the scale, and on the other someone who wants to clone a human being. Do you save the lives of hundreds of thousands of people, or stop everything for fear someone would abuse this [technology]? I'd prefer to help sick people.*
>
> Michael West, CEO of Advanced Cell Technology, quoted by W. Goldman Rohm in "Seven Days of Creation," Wired, January 2004

At this writing, the U.S. House of Representatives had passed a bill that bans all forms of cloning. The same measure was stalled in the Senate. In the meantime, Lanza pursues his work while it is still legal—and waits.

STEM CELLS MAY TREAT MUSCULAR DYSTROPHY

Studies in mice show that a type of stem cell in blood vessels could help patients suffering from the muscle-wasting disease, muscular dystrophy (MD).

Researchers in Milan and Rome have discovered that blood vessel stem cells actually cross from the bloodstream into muscular tissue, where they help generate new muscle fibers. It worked in mice with symptoms similar to those generated by MD, researchers say.

"Although these results are exciting, we have not cured the mice," said Giulio Cossu of the Stem Cell Institute of Milan, speaking at a press conference at the American Association for the Advancement of Science on July 10, 2003. "We believe this is a significant step toward therapy, but the question that keeps me awake at night is whether this will work in larger animals."

This particular kind of stem cell is new to scientists, having only been discovered a year ago. According to Cossu, they are still learning how to identify them by appearance and function, and they've so far only isolated them from fetal blood cells. Moreover, researchers need to refine the part of the procedure that involves inserting a healthy version of the gene that causes MD into the stem cell. Only more experimentation will show whether the procedure will ever be safe enough for humans.

As far as Cossu's mice go, they definitely improved as a result of the procedure. After treatment, their muscles were larger and had more muscle fibers. They were also able to walk on a wheel for a longer period of time than the untreated animals. "I'm convinced this is an important result, but this is still not the therapy—for the mice or for patients," Cossu told the press, underlining that the technique is still very much experimental.

It's a Fact

Fact: Almost all cloned animals are born by cesarean section, since they are too large to be born the traditional way. Scientists don't know why cloned animals often suffer from so-called large offspring syndrome. One theory says that it has something to do with the way genes are expressed during embryo development.

Fact: Embryonic stems are pluripotent. They have the potential to form any cell or tissue in the human body.

Fact: The American Medical Association in 2003 endorsed therapeutic cloning research, saying it is medically ethical. However, doctors opposing the practice may decline to perform it.

Fact: James Thomson of the University of Wisconsin at Madison was the first scientist to isolate human embryonic stem cells, launching a worldwide debate that still rages.

Fact: Think of a stem cell as a kind of master cell, an early-stage cell that has the potential to morph into any kind of cell in the body. A growing number of scientists believe that it will someday be possible to use cloning to generate stem cells to treat a wide range of human diseases, including Parkinson's, Alzheimer's, osteoporosis, and diabetes. Such ailments affect more than 125 million Americans.

CHAPTER NINE

GENE THERAPY

PROBABLY NO DNA science is at once as hopeful, controversial, hyped, and even as potentially dangerous as the discipline known as gene therapy.

In 1990, little Ashanti DeSilva, age four, went down in history as the first person in the world to be successfully treated with gene therapy. Ashanti suffered from severe combined immune deficiency syndrome (SCIDS), a single gene mutation that fairly crippled her immune system. As a result, she was prone to catching any passing bug. Children with this disease (known colloquially as "bubble boy" sickness) rarely survive to adulthood. But physicians—led by the University of Southern California's French Anderson—managed to actually insert DNA that would rectify the mutation that coded information for the defective protein behind the disease. To this day, Ashanti is apparently cured.

In 1999, a sadder kind of history was made. Eighteen-year-old Jesse Gelsinger volunteered for a gene therapy trial at the University of Pennsylvania relating to a chronic liver disease he'd been suffering from. He fell into a coma and died a few days later. Though others in the study had suffered few side effects, Gelsinger's body apparently had suffered an extreme immune response to the treatment. The Food and Drug Administration (FDA) immediately leaped into action, closing the study Gelsinger had participated in and putting in place ever-stricter controls on future gene therapy trials nationwide.

At this writing, gene therapy experiments are once again going strong at commercial and university labs around the world. To French Anderson, director of gene therapy at the University of Southern California Keck School of Medicine and the first scientist to conduct a gene therapy experiment in 1990, "Gene therapy and gene-based medicine will revolutionize medicine over the next ten to twenty years." The question isn't "if" gene therapy becomes a reality, Anderson said in a *Washington Post* article detailing the controversy, "The big question is when."[1]

How It Works

But before I get into "when," let's discuss the how and why. This is a therapy poorly understood by nonscientists. It is a prime example of one of those technologies that seems to have just suddenly materialized in the headlines of the public consciousness.

Gene therapy is an experimental technique that lets doctors treat a disorder by inserting new genes into a patient's cells. There are several possible approaches. Doctors may choose to replace a mutated gene with a healthy copy, as was the case with Ashanti DeSilva. They may choose to knock out, or inactivate, a mutated gene that isn't working correctly. Or they may elect to add a new gene entirely to help the patient's body to fight the disease, rather than replacing or knocking out an existing gene.

Recall that your DNA is located on chromosomes inside each of your cells' nuclei. Each cell is separated from other cells by its own cellular membrane. Also recall that, while each of your cells includes a complete copy of all of the DNA in your particular genome, exactly what genes are expressed (i.e., turned on) in that cell depends on exactly what kind of cell it is. That is, the genes expressed in a brain cell are going to be related to activity that cell needs to be a brain cell, and that will be different from the genes expressed in a stomach cell or a skin cell.

> *Gene therapy and gene-based medicine will revolutionize medicine over the next ten to twenty years. The big question is when.*
> *Gene therapy pioneer French Anderson, 2001*

Understanding that, you can easily see what some of the main problems are facing scientists who hope to perform gene therapy. Mainly, the problems are how to get a new gene into a cell, and how to make sure you hit the right cell.

Back in the late 1960s, scientists understood that creating stretches of genes in the lab would be possible. Nobel Prize–winning scientist Marshall Nirenberg in 1967 wrote about the feasibility of programming cells with man-made messages, and he discussed the promise and dangers that could result.[2]

But scientists also recognized the difficulty of getting the DNA through the membrane and actually incorporating it into a cell. Just injecting raw DNA into cells doesn't work too well. Scientists needed a method of actually getting the DNA into a cell before it was destroyed or ejected by the body's immune system.

That's where the modest virus comes in. A virus is the simplest organism there is—it is pretty much just genetic material wrapped up in a protein coat. And a virus can't live on its own—it survives and multiplies by parasitically attacking living cells and injecting its genetic material into cells. That makes it an ideal mechanism for getting genes into a cell. To make viruses work as a so-called vector (think of it as a "gene truck") that carries new genes into a cell, a scientist might take the damaging, infecting portion of the DNA out of the virus and add to it the desired gene segment. The fact that many viruses are cell specific (e.g., a certain virus may only infect heart cells or lung cells) helps scientists target just the desired cells. Typically, scientists use the relatively harmless adenovirus, or cold virus, as their vector, but methods vary. Some scientists have been turning to the retrovirus, a kind of virus that has RNA, rather than DNA, at its core.

> *We have created genetically engineered mice by adding genes to mouse embryos, so we know that the technique is ultimately practicable, though obviously a lot of safety issues have to be overcome. We therefore have to face the prospect that at some point, someone, somewhere—perhaps in twenty years—will cross the line and create a genetically engineered human embryo that will grow up to be a living human.*
>
> *Ethicist Lee Silver, 2003*

Another method of getting DNA into the cell is the so-called liposome method. This entails encasing new genes in a bubble of fat. In some studies, scientists found that fat-encased genes will melt into cells (which also are surrounded by a fatty membrane) in the same way that two bubbles merge into a single one. This approach does not work with most cells, though scientists have seen limited success in brain cells affected by Parkinson's disease and with skin cells where they attempted to treat some kinds of baldness.[3]

Once the vector, or gene truck, carrying the gene has been prepared, doctors usually inject it using a needle inserted into the targeted tissue of the body, where the correct cells ideally take it up. Alternatively, they may remove cells of the patient's body and mix them with the vector first, before reintroducing them into the patient.

Now that you understand how gene therapy works, let's talk about some specific successes and failures to get a handle on when it might be a feasible medical treatment for people who need it.

The Bleeding Edge

To treat little Ashanti DeSilva, French Anderson and colleagues at the National Institutes of Health (NIH) used an adenovirus (a modified cold virus) to carry a new copy of her mutated gene into her white blood cells. Because white blood cells don't live very long, Ashanti has since had to be retreated every few years. But she is living an otherwise normal life as a young teenager today. Without the treatment, odds are she would have by now been living in an enclosed bubble much like the SCIDS patient on whom the 1970s TV movie *Boy in the Plastic Bubble* was based.

But just as researchers got comfortable with the success of a treatment, the unexpected happened. At the Necker Children's Hospital in Paris, Dr. Alain Fischer treated ten children with SCIDS by inserting a new gene into their bone marrow. The majority of them appeared to be completely cured. But then two of the children contracted a rare form of leukemia. Everyone agreed it had to be more than just coincidence. In all likelihood, the virus had delivered the gene too close to an oncogene, a gene that controls cell growth. And in the process, it activated it, apparently causing growth to proceed out of control.

How could this happen? Bruce Sullenger, professor of surgery at Duke University, gave me a useful metaphor. He compares gene therapy to correcting a spelling error in a manuscript. If you just insert the correctly spelled version of a misspelled word at random in the document, the word is not always going to make sense. It could even confuse the meaning of another sentence.[4]

A ROCKY ROAD

Gene therapy has always made a lot of people nervous. In the early 1970s, when scientists first learned to clone DNA, public reaction was hostile. Opponents even managed to close down recombinant DNA experiments at Harvard and the Massachusetts Institute of Technology (MIT) for a few months. The fear was that a genetically engineered bacterium might escape.

Public fear diminished a bit after the National Institutes of Health got involved and, in 1974, formed the Recombinant DNA Advisory Committee (RAC, colloquially referred to as "the Rack") to be the watchdog on such safety issues. The NIH (through RAC) and the FDA are together in charge of approving all gene therapy studies. Additionally, universities are required to certify the safety of their experiments with the Institutional Review Board (IRB) working at that location.

Yet all the safeguards in the world won't matter if researchers choose not to adhere to them. A case in point: Martin Cline, a hematologist at the University of California, Los Angeles (UCLA) who went to Italy and Israel to perform his recombinant DNA experiments involving bone marrow treatments of patients with hereditary blood disorders. Cline never contacted the UCLA IRB to clear this experiment, and when this fact was leaked in a *Los Angeles Times* article in October 1980, heads rolled. UCLA officials forced Cline to resign from his post as department chair; he lost grants, and any time he applied for a grant afterward, it was attached to a report of his activities from 1979 to 1980. The fact that Cline didn't go to the review board, but rather, carried on like a maverick, was the key complaint against him.[5]

Cline's activities revived public concern about scientists "playing God." Groups including the United States Catholic

Conference, the Synagogue Council of America, and the National Council of Churches got a presidential commission involved. It, in turn, released a groundbreaking report, called *Splicing Life*, in 1982. The commission argued for the continuation of recombinant DNA research, saying that scientists are able to distinguish between what is acceptable gene therapy research and what is not. Also, the commission argued that the NIH's RAC add definitive ethical and social considerations to its long list of gene therapy concerns. In 1984, RAC created the Human Gene Therapy Subcommittee (HGTS) to do an initial review of gene therapy experiments, examining them from scientific, social, and ethical perspectives.

Through the HGTS, the government in 1990 approved French Anderson and Michael Blaese's gene therapy trials for children suffering from SCIDS. Ashanti DeSilva, the first person to receive FDA-approved gene therapy, was a direct beneficiary of this decision.

The 1999 death of teenager Jesse Gelsinger from treatment he received in the University of Pennsylvania gene therapy trial revived public scrutiny. In 2000, the U.S. Senate held hearings, raising serious questions about the effectiveness of government oversight.[6]

Gene therapy will succeed with time. And it is important that it does succeed, because no other area of medicine holds as much promise for providing cures for the many devastating diseases that now ravage humankind.

Gene therapy pioneer French Anderson

"It was an extremely unlikely event," Tony Blau, a University of Washington professor of medicine, told the *Seattle Post-Intelligencer* after the French children's leukemias were discovered. "But also exceptionally educational." Researchers are now trying to figure out why the inserted gene chose to integrate in that spot of all the places it could have integrated throughout the 3.1 billion-base-pair-long genome. "That's how nature teaches us."[7]

In 2004, researchers at Maryland's National Cancer Institute discovered something that may have cracked the case. When combined, a gene in the cold virus the French children were treated with and a gene involved in SCIDS caused the leukemia.

> *The hurdle now is in reducing all this science to commercial products.*
>
> Venture capitalist Gail Brown, quoted by David Shook in "Gene Therapy Is on the Mend," BusinessWeek, June 28, 2001

Lead researcher Utpal Dave said his work in mice shows that the leukemia suffered by the two SCIDS patients was rare, and that other forms of gene therapy likely won't carry the same risk.[8]

That may be the case, says the FDA's Phil Noguchi, but federal overseers still plan on keeping a close eye on gene therapy. Of the twenty-seven gene therapy trials it suspended after the French leukemia cases came to light, some have resumed. But all of them must conform to stringent government reporting guidelines.[9]

When Gene Therapy Is Fatal

The 1999 tragedy surrounding the fatal gene therapy of teenager Jesse Gelsinger was almost enough to stop the budding gene therapy revolution in its tracks. Gelsinger had suffered from a rare form of liver disease his entire life. The disease, called ornithine transcarbamylase deficiency (OTC), caused his liver to inadequately break down the chemical ammonia. Thanks to a low-protein diet and a medical regime of thirty-two pills a day, he was living a fairly normal and healthy life when he entered a voluntary gene therapy study at the University of Pennsylvania. The therapy wouldn't help him directly, he knew. It was designed to test the safety of a therapy for newborns who suffered from his disorder.

His father, Paul Gelsinger, reviewed the program and encouraged his son to participate. He told a PBS audience that he and his son had no worries about it. "Jesse was doing exceptionally well on his medications, and nothing should have prevented him from living a full and happy life. He believed, after discussions with representatives from Penn, that the worst that could happen in the trial would be that he would have flu-like symptoms for a week. He was excited to help."[10] Like the other sixteen patients in the trial, Jesse Gelsinger was injected with an adenovirus carrying a copy of the ornithine transcarboxylase (OTC) gene to replace the one not functioning in his liver cells. The mixture was delivered directly into the hepatic artery leading to his liver.

"Less than twenty-four hours after they injected Jesse with the vector in the amount that only one other person had ever been given, Jesse's entire body [reacted] adversely," says Paul Gelsinger. "He went into a coma before I could get to Philadelphia and see him, and died two days after my arrival, directly as a result of that gene therapy experiment."[11]

> People have to understand that this has really never been done before. Gene therapy uses several types of protocols that are different from what you normally have for a drug. That's because, with gene therapy, it's the cells in the body [that] make the final therapeutic compound.
>
> John Monahan, CEO of Avigen, Inc., in "Gene therapy Is on the Mend," BusinessWeek, June 28, 2001

To this day, no one is exactly certain why Gelsinger's body reacted as it did. But it is clear that his immune system launched a venomous attack against the vector, causing a fatal cascade of events, beginning with organ failure and coma and leading eventually to his death.

In the FDA investigation that ensued, several items came to light. One was that Gelsinger's ammonia levels were too high to qualify him for the study in the first place. Also, university researchers had failed to disclose some key information—namely, that two patients had experienced severe side effects in a previous trial, and monkeys had died during the university's animal experiments with the procedure.[12]

"As tragic as [the Gelsinger] event was, I think it helped people understand how to move forward in the future safely and carefully," says Paul Fischer, CEO of the gene therapy company GenVec, Inc. "It made people double-check the safety issue."[13]

Gene Therapy in the Labs

Researchers around the world now are taking a renewed look at gene therapy trials. In the United States alone, there are no fewer than a dozen companies plowing full steam ahead. "We do seem to have turned the corner," says Anderson, "and there are a number of clinical trials that are starting to show success."[14]

> We've got to keep at it. It would be easy to say this is too hard. This is too exciting an odyssey to miss.
>
> Barrie Carter, Targeted Genetics chief scientific officer, quoted by Carol Smith in "Seattle Home to Cutting-Edge Gene Therapy Research," Seattle Post-Intelligencer, February 28, 2002

One of them is GenVec's. The Gaithersburg, Maryland, company's BioBypass product is targeting coronary artery disease (CAD) and peripheral vascular disease (PVD), conditions caused by blocked arteries that slow blood flow to the heart and legs. CAD and PVD affect millions of Americans. BioBypass, which at this writing is in late-stage phase II clinical trials, injects patients with new genes. These genes help the body grow new blood vessels, enabling blood to go around the clogged arteries altogether.

Another interesting project is ongoing at UCLA. There, researchers led by William Pardridge have been working with monkeys on a potential gene therapy treatment for the neurological disorder, Parkinson's disease. Getting genes into the brain has been near impossible because viral vectors are physically too large to pass through the brain membrane.

"This is a monumental problem for drug development," says Pardridge, because 98 percent of available intravenous drugs and 100 percent of oral medications cannot get from the blood to the brain.[15]

His solution: Create a "molecular Trojan horse" to coat the new genes with lipids, and coat that with a chemical called propylene glycol (PEG) that keeps the coated genes from being absorbed by the liver and other tissues. Then, they can slip right from the blood and directly into the brain.

TOP PUBLIC COMPANIES DOING GENE THERAPY

1. Targeted Genetics Corp. (Nasdaq:TGEN): Molecular medicines

2. Introgen Therapeutics, Inc. (Nasdaq: INGN): Gene therapy for cancer

3. Valentis, Inc. (Nasdaq:VLTS): Cardiovascular therapeutics

4. GenVec, Inc. (Nasdaq: GNVC): Biopharmaceuticals for cancer, heart disease, and vision loss

5. Cell Genesys, Inc. (Nasdaq: CEGE): Cancer vaccines and gene therapies

6. Avigen, Inc. (Nasdaq: AVGN): Gene therapy for chronic diseases

7. Vical Incorporated (Nasdaq: VICL): Cancer therapies

8. Onyx Pharmaceuticals (Nasdaq: ONXX): Cancer therapies

In a 2003 issue of the newsletter of Children's Neurological Solutions, Eain Cornford, a professor of neurology at UCLA, was effusive. "This particular work is the most innovative and potentially the most groundbreaking that anybody in blood-brain research has done . . . it could well revolutionize the field."[16]

Other gene therapy trials are subtle twists on the idea. At the University of North Carolina, researchers led by Ryszard Kole are hard at work on a therapy for thalassemia, a form of anemia that is the world's most common single-gene disorder. More than 100,000 children are born with it every year, and most often they are of Mediterranean and Southeast Asian descent. (These days, most pregnant mothers can be screened for thalassemia on request.) The disorder involves one or two mutations in the genes that code for the production of hemoglobin; as a result, sufferers can't even make the slightest exertion.

Rather than trying to replace the defective genes, Kole and his researchers are working on repairing the defective RNA that emanates from the cell. "This approach is a lot more straightforward than conventional gene therapy," Kole says, adding that conventional gene therapy is more difficult because researchers don't have the ability to control the activity of any specific gene.

"But by repairing messenger RNA rather than trying to replace a damaged gene, you are using the cell's own regulatory mechanisms to produce normal hemoglobin in the correct quantities," he says.[17] Basically, Kole's experiment tricked the body's machinery for making red blood cells into producing normal hemoglobin. Experts say even a small improvement in the production of normal hemoglobin will make a huge difference to these patients.[18]

The list of other promising take-offs on gene therapy experiments is long. Beverly Davidson and her team at the University of Iowa are working on gene therapy for Huntington's disease, not by adding a new copy of a defective gene, but by "silencing" one of a patient's genes responsible for the disease. (Huntington's chorea patients have a mutation on chromosome 4, with too many repeats of the triplet CAG, leading to an overproduction of glutamic acid, which eventually kills off part of the brain.)

Davidson's team accomplished this process in mice by using a natural cell technique for fighting infections, called RNA interference. RNA interference is an exciting technique in biology. Essentially, instead of trying to change a given gene, researchers attempt to block the action of the RNA that "reads" the gene to make protein. If the RNA is unable to read the bases to figure out which proteins to make, essentially the gene has become inactive.

"When I first heard of this work, it just took my breath away," says Nancy Wexler of Columbia University Medical School, who is president of the Hereditary Disease Foundation in New York.[19]

Stocks of [gene therapy] companies may be a little subdued, because the scientific progress is not moving forward very fast. Quite a bit of research still needs to be done. But this remains a very promising field for disease research.

George Wideru, senior scientist at Genetronics Inc., quoted by David Shook in "Gene Therapy Is on the Mend," BusinessWeek, June 28, 2001

Some experiments serve as important reminders that gene therapy is considerably harder than expected. Consider sickle cell disease (SCD), also known as sickle cell anemia. It is a straightforward and well-understood genetic disorder. Prevalent among those of African descent, a single-gene mutation causes abnormal (and abnormally sickle-shaped) hemoglobin. "Everybody thought it would be the first genetic disorder cured by gene therapy, that it would be simple, but it turned out to be completely different. It was a real challenge," says Philippe Leboulch, an MIT and Harvard gene therapist who has worked with SCD for more than a decade.

It wasn't until fairly recently that Leboulch and his team got a handle on the disease, however. Using a gene truck, or viral vector, derived from HIV, they were able to inject SCD mice with a new gene. That gene was for hemoglobin production. And it worked. The great majority of the mice remain cured more than a year later.

> *The Cassandras and Jeremiahs and Gloomy Guses are a part of the problem, not of its solution. Technology, whether of the "hard" physical kind or the "soft" biological kind, is man's creation and man's hallmark. . . . To be civilized is to be artificial, and to object that something is artificial only condemns it in the eyes of subrational nature lovers or natural-law mystics.*
>
> Ethicist C. John Fletcher, writing with W. French Anderson in "Germline Gene Therapy: A New Stage of Debate," Law, Medicine & Health Care, v. 20, 1992, pp. 1–2

Gene therapy pioneer French Anderson summed it all up in an article he wrote for *Science*: "The field of gene therapy has been criticized for promising too much and providing too little during its first ten years of existence. But gene therapy, like every other major new technology, takes time to develop. Antibiotics, monoclonal antibodies [antibodies designed by researchers to attack one kind of a cell, say, a tumor cell], organ transplants, to name just a few areas of medicine, have taken many years to mature.

"Major new technologies in every field, such as the manned rocket to the moon, had failures and disappointments," wrote Anderson. "Early hopes are always frustrated by the many incremental steps necessary to produce 'success.' Gene therapy will succeed with time. And it is important that it does succeed, because no other area of medicine holds as much promise for providing cures for the many devastating diseases that now ravage humankind."[20]

¿ WHEN GENE THERAPY MAKES SENSE

Gene therapy trials show the greatest success when applied to diseases resulting from single gene defects. The FDA has approved gene therapy trials for severe combined immune deficiency, cystic fibrosis, Gaucher's disease, hypercholesterolemia, and Huntington's chorea. Other labs are looking at gene therapy to subdue forms of cancer, heart disease, Parkinson's disease, and Alzheimer's.

It makes the most sense to try gene therapy, experts say, when you are dealing with an incurable, life-threatening illness. In such cases, when you know the cell types affected by the disease, the defective gene has been isolated and it is possible to safely introduce a new gene.

Thanks to the Human Genome Project, researchers are gaining rapid knowledge of how genes and their associated proteins contribute to disease. The list of gene therapy trials is expected to grow rapidly over the next few years.

...It may...mark the end of human life as we and all other humans have known it. It is possible that the nonhuman life which may take our place will be superior, but I think it most unlikely and certainly not demonstrable.

Ethicist Leon Kass, in "Germ-Line Gene Therapy: A New Stage of Debate," Law, Medicine & Health Care, *1992*

Gene Therapy for the Ages

Until now, we've been talking solely about so-called somatic gene therapy, or gene therapy that applies to genes on a patient's nonsex chromosomes, the autosomes. In somatic gene therapy, while you are aiming to cure the patient by replacing or knocking out defective genes, you are not permanently improving the genetic makeup that individuals pass along to their children through their sex cells— that is, their sperm and eggs.

Manipulating the genes on the sex cells—or what is called germ-line gene therapy—is decidedly more controversial. It is easy to see why. You are essentially modifying the human race—or one ancestral line, anyway. Though the U.S. government currently bans federal funding for germ-line gene therapy, it is easy to enumerate its benefits. For one thing, it may be the only effective method of tackling some genetic diseases. Because it actually modifies the cells passing down DNA during reproduction, the likelihood of future generations suffering from a given genetic disease is virtually eliminated.

> *One cannot see anything intrinsically forbidden or evil in gene therapy, whether somatic or germ-line. Infinite possibilities of power are open to humanity. The ethical problem is not in the acquisition of this power, but in its wise use.*
> Gregorius, Greek Orthodox Bishop of Delhi

The arguments against germ-line gene therapy are compelling. There is, of course, the familiar slippery-slope debate. If we begin treating disease at the germ-line level, where do we stop? Is there a gray area where a "defect" is treated as a "disease"? The fear is that at some point, gene therapy could open the floodgates to attempts at genetically modifying people for traits that are not at all disease related. Should we eliminate deaf people, for instance, or people with widow's peaks?

"I think we should be very careful about this," says geneticist Steve Jones. "Anatomy began in the sixteenth century, yet it took us another 400 years to carry out the first transplants. Today, we have only just begun to isolate genes. We should not expect to be able to transplant them overnight. In any case, if you really want to engineer your child's IQ, stick to the old ways. Send them to Eton [College]. And if governments want to improve the nation's intelligence, the best value for money would be to double teachers' salaries."[21]

GENE THERAPY AND RADIATION: A PROMISING COMBINATION

Nearly 30,000 men will die this year from prostate cancer, the leading cause of death among males after lung cancer, according to the American Cancer Society. But a new therapy that combines gene therapy techniques with radiation offers some hope. "Our belief is that gene therapy could make conventional cancer therapies such as radiation therapy more effective," says Svend Freytag, the lead author of a study at the Henry Ford Health System in Detroit.[22]

Essentially, the new approach entails injecting the adenovirus (cold virus) as the gene truck (viral vector) carrying new genes into the prostate. It spreads, infects, and apparently weakens tumor cells, in addition to carrying the genes to a wider group of prostate cells. Then, researchers bombard the tumor cells with radiation.

The combination looks effective. In the study, researchers treated fifteen men with advanced forms of prostate cancer. All of them almost immediately showed a decrease in levels of the prostate-specific antigen (PSA) protein, a common marker for prostate cancer. And a year later, ten of the men were entirely cancer free.

Researchers intend to immediately expand the trials. In early 2004, Freytag received a $9 million cash grant to commercialize his efforts.

In wanting to become more than we are, and in sometimes acting as if we were already superhuman or divine, we risk despising what we are and neglecting what we have.

"Beyond Therapy: Biotechnology and the Pursuit of Happiness," a report of the President's Council on Bioethics, October 2003

There are practical considerations, too. If you are going to modify the germ line of generations, you owe it to the patients and their ancestors to continually follow up. After all, the long-term effects of this therapy are unknown. Also, how do you attach a price to the procedure? Gene therapy is expensive and will likely remain so. Are you intentionally creating a genetic underclass whose ancestors could not afford the technique?

And what of the rights of the fetus, if any? Obviously, a fetus cannot agree to a procedure that changes its very genetic makeup. The same argument applies, incidentally, to more traditional, somatic gene therapy, which is often performed on babies. Is it acceptable to proceed with gene therapy on a patient, even if a patent is too young to understand how she is being treated?

> *Progress in gene therapy has admittedly been slow in the early period. But it will accelerate. Too much hope is at stake, and too much venture capital poised, to permit failure. Once established as a practical technology, gene therapy will become a commercial juggernaut.*
>
> Biologist Edward O. Wilson in Consilience, (New York: Knopf, 1998), p. 301

The issues surrounding gene therapy will only heat up as we move toward the future. "We have created genetically engineered mice by adding genes to mouse embryos, so we know that the technique is ultimately practicable, though obviously a lot of safety issues have to be overcome," says Princeton ethicist Lee Silver. "We therefore have to face the prospect that at some point, someone, somewhere—perhaps in twenty years—will cross the line and create a genetically engineered human embryo that will grow up to be a living human."[23]

James Watson, co-discoverer of the double helix, has quite a different view about gene therapy on the sex cells. "I'm in favor of going forward, though most of my fellow scientists say they are against it. I believe they don't want to alarm the public by possibilities which will never exist. On the other hand, by adding appropriate genes, we already can improve the memory of mice. Why not the same with humans? To me, it's common sense to take steps which might give our future descendants more effective brains. I don't see who we're truly offending by trying to enhance ourselves. To me, it goes against human nature that people should not try to improve the lives of their children and those that follow."[24]

It's a Fact

Fact: Gene therapy is still experimental. Researchers are running hundreds of clinical trials worldwide to determine how it can help cure cancer and other complex diseases.

Fact: In gene therapy, doctors generally try to replace missing or altered genes with healthy ones. Instead of getting a drug, a patient gets a new gene that alters the genetic makeup of her cells. In theory, this will be particularly effective in diseases involving mutations on just one gene, such as hemophilia and cystic fibrosis.

Fact: Another promising use of gene therapy is to improve the way cells function. For instance, doctors might add genes that stimulate the immune system to attack cancer cells, or resist human immunodeficiency virus (HIV) or acquired immunodeficiency syndrome (AIDS).

Fact: Generally, a gene that is inserted directly into a cell doesn't function. Rather, it must be carried into the cell using a delivery mechanism called a vector. The most common vector in gene therapy trials is the virus, which has a natural ability to enter a cell's DNA. Happily, viruses are also fairly easy to disable so that they don't reproduce themselves.

Fact: The difference between a virus and a retrovirus is that the latter contains ribonucleic acid (RNA) as its genetic material instead of DNA. Retroviruses produce an enzyme called reverse transcriptase, so they can transform their RNA into DNA, which then becomes part of the host cell's DNA.

Fact: Getting a gene therapy trial approved isn't easy. First, at least two review boards must approve it at the institution or university where the scientist works. Then, the FDA, which regulates all gene therapy products, needs to give the go-ahead. Finally, any trial that is funded by the National Institutes of Health (NIH) needs to be registered at the NIH's Recombinant DNA Advisory Committee (RAC), often called "the Rack."

Fact: Why the nervousness about germ-line gene therapy? Because it would forever change the genetic makeup of everyone in that individual's family tree down the road. That means it would affect the human gene pool. Germ-line therapy modifies the X and Y chromosomes. Even though, theoretically, it would only be performed to improve genes—for instance, to remove a mutation that causes a hereditary disease—errors or mistakes in judgment would have long-term consequences. The NIH does not currently approve any germ-line gene therapy experiments.

Fact: Genetic enhancement is still mostly science fiction now, but ethicists worry that if it becomes easy, it would become available only to the wealthy. In effect, that would create a kind of genetic underclass and redefine what "normal" means. For instance, people with just average intelligence would be considered subnormal if the well-off could engineer their offspring to be smarter.

CHAPTER TEN

DNA AND SOCIETY

SO FAR, I've tried to give you a clear, concise look at where various cutting-edge DNA sciences are headed. I've covered DNA fingerprinting and how it is exonerating death row inmates and other convicts. I've outlined the exploding potential of prenatal and adult gene testing, not to mention the DNA testing that is solving mysteries that are in some cases hundreds of years old. I've covered what's happening in the world's most progressive labs in anti-aging research, cancer treatments, gene therapy, even the touchy issues of stem cell technology and cloning.

To stop now would be irresponsible. I can't fly you over such a broad landscape of DNA issues and avoid what many consider to be the primary question—namely, where will all this new knowledge lead us?

What will it mean to our lives, our families, our society, and the society of future generations? Yes, we are the first life-forms on the planet to be able to look at our own recipe. We are learning what we are made from, and now we are quickly moving to the next step of finding out what it all means and, from there, how to manipulate what it means. There will, of course, be consequences, negative and positive and in between. But what consequences?

Should you be worried?

At the outset, I advised that you would need to understand the basics of DNA technology to keep abreast of the phenomenal change all around us, whether you just want to follow the DNA sciences in the news or hope to actually invest in biotechnology companies. Now, I'll add to that and say that you—as a human being and as a world citizen—fairly owe it to the rest of us to understand the questions and controversies surrounding the DNA sciences if you want to have a voice in what is to come. It is a heavy thought, but it is true. If you choose, you can shape the technology that's coming, rather than choosing to let it shape you when it gets here.

Harold Varmus, former director of the National Institutes of Health (NIH), has noted that elected officials—many of them clueless about the science and the issues—are increasingly making decisions about what happens to DNA information once it is collected. Those questions deal with whether insurance companies can take a look, whether the police can forcibly demand it, and a myriad long list of other issues that will affect all of us this century.

Citizens especially need to understand these issues in order to take a stand on them. We need to switch a light on if we are to anticipate what's coming up ahead.

The Old Eugenics

No one truly saw the Holocaust coming. When the Nazi death camps were finally revealed to a horrified world, the world was uniformly shocked. Yet eugenics, the pseudoscientific "master race" movement that the Nazis based most of their theories upon, was alive and thriving in the United States for at least a quarter century prior to World War II.

By the time Hitler took power and began applying his eugenics theories to Europe's masses, twenty-four U.S. states had forcible sterilization laws designed to keep the "unfit" and "feeble-minded" from bearing children, all in the interest of improving the American gene pool.

> *We have seen more than once that the public welfare may call upon the best citizens for their lives. It would be strange if it could not call upon those who already sap the strength of the state for these lesser sacrifices, often not felt to be such by those concerned, in order to prevent our being swamped with incompetence. It is better for all the world, if instead of waiting to execute degenerate offspring for crime, or to let them starve for their imbecility, society can prevent those who are manifestly unfit from continuing their kind. The principle that sustains compulsory vaccination is broad enough to cover cutting the Fallopian tubes. . . . Three generations of imbeciles is enough.*
>
> U.S. Supreme Court Justice Oliver Wendell Holmes, writing in the Buck v. Bell *decision, 1927*

Charles Darwin's cousin, Francis Galton, is generally blamed for coining the term *eugenics* from the Greek term for "well-born." (Galton is better known for his invention of human fingerprinting as a means of identifying people.) As early as the 1880s, Galton promoted the concept of improving the human race by making sure the most talented and attractive men mate with the most talented and attractive women, while at the same time limiting the reproductive potential of the not-so-fortunate. It was a natural extension, he wrote, from the concepts of natural selection to a purposeful improvement of the human race. "What nature does blindly, slowly, and ruthlessly, man may do providently, quickly, and kindly," he wrote.[1]

Galton's theories were rediscovered along with the Mendelian principles of inheritance at the turn of the twentieth century. The theory was that if you could breed peas for height, seed texture, and leaf color, why shouldn't you attempt to breed people for beauty, brains, character, and courage? This is how genetics made its dubious debut in the public consciousness, veiled by eugenicists as a science for the greater good.

> *There is no great invention, from fire to flying, that has not been hailed as an insult to some God.*
>
> Biochemist J.B.S. Haldane in Daedalus, or Science and the Future (New York: Dutton, 1923)

It would've been just another half-baked fringe theory, but doctors, scientists, politicians, and powerful and wealthy foundations such as the Rockefeller Foundation and the Carnegie Institution got firmly behind it. Suddenly, books and articles spouting the new "race science" were everywhere. At the Kansas Free Fair of 1924, notes Yale ethicist Daniel Kevles, alongside the prize pig and pumpkin contests there was a "human stock" competition for grade A families. The prize was a "Governor's Fitter Family trophy," and winners could "perform" in any one of three categories: small, medium, and large.

But eugenicists didn't just encourage the "best" to procreate. Soon, they began focusing their efforts on reducing the American population of "misfits and mongrels," the perceived worst of society. There were marriage prohibitions, human breeding programs, and, finally, the passage of sterilization laws in states across the country.

In all, the United States forcibly sterilized more than 60,000 people (many of them women considered "wanton" and "wild") in the hopes of cleaning up the American gene pool.

Edwin Black, in his book *War Against the Weak*, focuses on that sweeping, devastating effort. "The victims of eugenics were poor urban dwellers and 'white trash' from New England to California," he writes, "immigrants from across Europe, Blacks, Jews, Mexicans, Native Americans, epileptics, alcoholics, petty criminals, the mentally ill, and anyone else who did not resemble the blonde and blue-eyed Nordic ideal the eugenics movement glorified."[2]

By an eight to one decision, even the U.S. Supreme Court stood firmly behind eugenics, upholding Virginia's eugenic sterilization law and the forcible sterilization of a young unwed mother. Justice Oliver Wendell Holmes wrote the opinion, which is startling in retrospective:

> It is better for all the world, if instead of waiting to execute degenerate offspring for crime, or to let them starve for their imbecility, society can prevent those who are manifestly unfit from continuing their kind. The principle that sustains compulsory vaccination is broad enough to cover cutting the Fallopian tubes. Three generations of imbeciles [a reference to the young woman, her two-month-old child, and her mother] are enough.

Three years before Holmes's written opinion, Adolf Hitler admitted fascination with the course eugenics was running in the United States. In 1924, in *Mein Kampf*, Hitler wrote of his admirations for America's newly toughened immigration laws and other eugenic policies. "There is today one state," he wrote, "in which at least weak beginnings toward a better conception [of immigration] are noticeable. Of course, it is not our model German republic. It is the United States."

Hitler told a colleague, years later, "I have studied with interest the laws of several American states concerning prevention of reproduction by people whose progeny would, in all probability, be of no value or be injurious to the racial stock."

The eugenics movement was by no means confined to the United States and Germany. It flourished in Canada, Britain, and Scandinavia as well. Sweden sterilized upwards of 50,000 women, most of them suffering from genetic diseases and other disorders.

Eugenics was, of course, discredited after the Nazis death camps and the grisly experiments on twins performed by Josef Mengele came to light. (The Nazis facing trial for war crimes used Oliver Wendell Holmes's writings as a defense at one point.) The fear nonetheless remains in many critics' minds about a new kind of eugenics, a softer variety that may result from our newly gained knowledge of individual genetic mutations and a similar willingness to run down human rights in the interest of the public good . . . or the private profit.

"In the public realm," writes Kevles, "as the costs of medical care continue to rise, the increasing acquisition of genetic information could conceivably lead to a renewal of the ethical premises of the original eugenics movement, an insistence that the reproductive rights of individuals must give way to the medical-economic welfare of the community as a whole."[3]

The concept of the natural does not deter science. In fact, I would put it the other way. To do battle with the natural seems to be a critical driving force. Science has a long and deep history of completely disrespecting the concept of the natural.

David J. Rothman, "Redesigning the Self," in The Genomic Revolution, (Washington D.C., Joseph Henry Press, 2002)

"As a country, we have not outgrown bigotry, nor our regular desire to find scapegoats for economic conditions, nor the need to enlist science as the panacea for social conditions," adds attorney Paul Lombardo, who has served as director of the Center for Biomedical Ethics at the University of Virginia. "The current hype that surrounds genetics will provide plenty of fuel for those who wish to push neo-eugenic schemes, whether or not they use the discredited description of 'eugenics,'" Lombardo says.[4]

A New Eugenics?

In *War Against the Weak*, Edwin Black points out that if new eugenics were to arise out of our new knowledge of DNA, it would start by creating a genetic underclass—"an uninsurable, unemployable, and unfinanceable underclass." And to do that, society must systematically categorize everyone's DNA so it can separate the wheat from the chaff. Black believes "the process has already started."

> *We are on the verge of a true evolution in medicine. But there is a chance it will not happen as we hope. It will not be a failure of the science. There is increasing evidence [that] people fear their genetic information will be used to deny them health insurance or a job. This fear is keeping them from seeking medical help. The revolution at hand may not be realized because people are afraid to take part in it.*
>
> Craig Venter, founder of Celera Genomics

The FBI is already collecting DNA samples from any U.S. citizen convicted of a crime, and some states are getting samples from those arrested (and presumed innocent, by law). The Department of Defense collects samples from all military personnel, and the federal government collects samples from many civil employees. And then there are the infant blood samples stored from the heel-sticks doctors use to test newborns for blood type and the disorder PKU. Stored as "Guthrie blots," California alone retains more than seven million samples. Combine that with the routine blood tests and occasional genetic tests many people get, and you can see that we are well on our way to a situation where every American's DNA is in a databank, somewhere.

"What is important for the public to decide is, 'Do we want the government to have our life code, the code that makes us who we are?'" says Steven Benjamin, chairman of the Forensic Evidence Committee of the National Association of Criminal Defense Lawyers.[5]

Quick and Dirty Answers

"I am concerned about eugenics with a 21st-century twist," says Troy Duster, the New York University sociologist who chaired the national advisory committee of the Human Genome Project's Ethical, Legal, and Social Implications program in the late 1990s.

Duster told me he believes that soon, potentially everyone's DNA will be stored in a DNA databank somewhere. The FBI collects the DNA of anyone convicted of a crime (and sometimes those just accused of them) for its huge and rapidly growing Combined DNA Index System (CODIS) national database.

In addition to fears of civil rights violations—such as forcibly taking DNA from suspects, the accused, and parolees—Duster says there's a chance criminologists might start correlating single nucleotide polymorphism (SNP) variations in the human genome to criminal potential. For instance, with the majority of prison inmates being African-American, you would expect to see a higher than average rate of sickle cell anemia there, in the prison population, too. But it would be wrong—and dangerous—to correlate SNPs like the one that causes sickle disease with a propensity for crime.

> *Most people think it's a good thing to have DNA in databanks because it is helping you to catch criminals. Very few people are thinking of it as a potential abuse of power.*
>
> *Sociologist Troy Duster*

"This is where the danger is," says Duster, "it's a false sense of precision, where you think that you are getting answers to problems that are actually more complex. You think that DNA is going to help us solve certain problems, when in fact it just may deliver us quick and dirty [and wrong] answers."[6]

Another thing to think about, says ethicist Philip Bereano, is that a genetic test may indicate a mutation, but it doesn't indicate an absolute likelihood that a person will suffer from a certain genetic disease.

Consequently, genetic tests that are given to people, forced or otherwise, will often reveal information that is by no means absolute.

Genetic determinism—the doctrine of categorizing people solely on their genetic merits or problems—"is so often based on crap," Bereano told me. "And there is some of that going on now. Eugenics, unfortunately, is the logical result of genetic determinism. The thought that follows 'these things are bad' is, 'let's use technology to fix them.' But we have a long history that shows us what the problems of this approach are."[7]

> *You can't help but worry that as we get more powerful we'll start manipulating human beings in ways that will be very difficult to accept and potentially very dangerous.*
>
> David Baltimore, Caltech president

Forced Genetic Testing in the Workplace

In these early days of genetic testing, there are already glaring examples of employers testing employees' DNA without their knowledge.

"It's all about power," says Bereano. "These employers want the power of surveillance, even in a situation when it is not rational."

A case in point involves Lawrence Berkeley National Laboratory (LBNL). This California Department of Energy lab gave routine medical exams to its employees, but employees didn't know that if they were black or female, they were getting additional tests, too. LBNL was testing women for pregnancy and African-Americans for the presence of the gene signaling susceptibility for sickle cell anemia. It was even reportedly testing African-American and Hispanic employees for syphilis.

When they found out, employees sued. And at the appellate court level, they won. In 1998, the Ninth Circuit Court of Appeals ruled that the company certainly violated the employees' privacy. Judge Stephen Reinhardt, in the opinion, noted that "the conditions tested for were aspects of one's health in which one enjoys the highest expectations of privacy." The employees won $2.2 million, and their data was deleted from the computers.

In 2002, Burlington Santa Fe, a major North American railroad company, paid the same amount in settlement. The company was also testing employees without their knowledge, but this time in an attempt to determine if employees had an increased susceptibility to carpal tunnel syndrome. (In court documents, it remained unclear whether executives were misled, because the mutation they were testing for may be unrelated to carpal tunnel.)

In court, Burlington managers said they were simply trying to limit their insurance liability.

> *Random throws of genetic dice take away from too many infants the chance to fully participate in human life.*
>
> James Watson, remarks to the Commonwealth Club of San Francisco, October 9, 2003

Testing and Discrimination

Would you receive genetic testing if you feared your employer or insurance company could use the results against you? Scientist Craig Venter bets you wouldn't.

Venter testified before a congressional subcommittee in an attempt to convince legislators not to let insurance companies discriminate on the basis of genetic testing.

"We are on the verge of a true evolution in medicine," Venter testified on behalf of the Biotechnology Industry Organization (BIO). "But there is a chance it will not happen as we hope. It will not be a failure of the science. There is increasing evidence people fear their genetic information will be used to deny them health insurance or a job. This fear is keeping them from seeking medical help.... The revolution at hand may not be realized because people are afraid to take part in it."[8]

"Consumers are very worried about privacy and whose hand this medical and genetic information will fall into," said Nancye Buelow, vice president for consumers at Genetic Alliance, a coalition of groups that claims to offer "safe harbor" to people whose lives have been affected by genetic findings. In the trade publication *Risk & Insurance*, she said, "There are people denied insurance because of genetic conditions and people who have had insurance canceled or fees raised. So there is cause for concern on the consumer side."[9]

> Our own DNA belongs to each of us as individuals. No one should have the right to examine it without our consent. In saying this, I know that the matter will prove more complex than we like. With time, we will know enough about our genes to be able to predict, say, whether we will likely have a long life. In which case we will probably decide not to take out large insurance policies at an early age. Insurance companies will know that we have this ability, and so they will demand the right to look at our DNA, say, if we wanted to take out a $10 million policy. Some compromise must be found; we don't want insurance companies to go out of business.
>
> James Watson, address to the Commonwealth Club of San Francisco, October 9, 2003

While insurance companies lobby Congress not to pass laws that would keep them from obtaining genetic information, claiming the fears are unfounded, ethicist Bereano says he has found abounding anecdotal evidence to the contrary. He says he's received reports that a health maintenance organization (HMO) reportedly told a pregnant woman whose fetus tested positive for cystic fibrosis that the HMO would cover the cost of an abortion, but not the cost if the woman carried the baby to term. A young boy, who carried a gene causing a susceptibility to a heart disorder, was denied insurance, even though he was on medication that removed his risk. A healthy woman mentioned in passing to her family doctor that her father had Huntington's disease. An insurance company found a note the doctor had written about it and rejected her disability insurance.[10]

> *Whenever a child was born, it was taken to a council of elders for examination. If the baby was in any way defective, the elders dropped it into a chasm. Such a child, in the opinion of the Spartans, should not be permitted to live. Newborn children were washed with wine so they would be strong. They grew up free and active, and without any sort of crybaby ways. Spartan children were not afraid of the dark, or finicky about their food.*
>
> Plutarch, describing the ancient Sparta led by Lycurgus, 800 B.C.

Though several bills have circulated Congress to prevent insurers and employers from genetic discrimination, at this writing, nothing has been signed into law.

"Your video rental records are more protected" than your genetic privacy is, says Joanne Hustead, a director at the National Partnership for Women and Families, who has been urging federal legislation. "There is no federal law on the books to protect [private sector] employees because members of Congress have their heads in the sand," she says.[11]

Genetic Testing and Societal Ills

Widespread DNA sampling and genetic testing could also lead to society's hiding its head in the sand, worries Helen Wallace, who leads the GeneWatch citizen watchdog group in the United Kingdom. Left unchecked, it could lead to a lazier society, one less inclined to fix social ills when it can more easily point at the genetic causes behind them.

"The adverse health impacts of smoking, poor diets, poverty, and pollution are not limited to individuals with bad genes," she explains.

"The current massive increase in obesity, for example, is not caused by an increase in genes for obesity. So, although we all vary slightly in our biology, and some of this variation is genetic, the big changes in the incidence of common diseases [in time, or between different countries] are due to social, economic, and environmental factors."

The genetic approach to "prediction and prevention" of these diseases is a distraction from the risk factors that can be changed—environmental ones, in the broadest sense—to the ones that can't be changed—our genes. This is sometimes useful for diseases where the risk is dominated by a single genetic factor, but it does not make sense in most cases.

Focusing on genes for obesity or cancer, Wallace adds, "just shifts the blame from . . . products or pollution to the individual with 'bad genes' and implies that policies restricting consumption or pollution can be applied to only a minority of individuals. It's a bit like saying, 'Let's try to find the people who are genetically susceptible to getting run over,' rather than providing a safe place to cross the road or tackling speeding."[12]

> There is today one state, in which at least weak beginnings toward a better conception [of immigration] are noticeable. Of course, it is not our model German republic. It is the United States.
>
> Adolf Hitler, Mein Kampf

The potential for new age health scams and rip-offs is enormous, she adds. If cheap genetic tests are available over the counter, so too will over-the-counter drugs and supplements be available, targeting what Wallace calls "the healthy ill."

"The marketing strategy might also involve other 'individualized' products supposedly tailored to your unique genetic makeup: skin creams, vitamin regimes, and ultimately 'functional foods,' perhaps genetically engineered to supposedly contain the right vitamins for you. Some of these products may cause harmful side effects.

Others will just be a rip-off for customers. None are likely to make big improvements to health," Wallace says.[13]

Affording Genetic Upgrades

If there is a new eugenics movement, as many ethicists have pointed out, it won't target a certain race or religion. It will favor the genetically advantaged over the genetically disadvantaged. And the wealthiest of society will be able to afford to place themselves in the correct category.

> *The state is saying, in effect, you may be a danger in the future because you were in the past, and therefore we need to register your DNA. That is a fundamentally different way than the government has heretofore been permitted to treat its citizens.*
> Defense attorney Benjamin Keehn, in an interview on PBS's NewsHour, July 10, 1998

The poor, after all, currently have the least access to adequate medical care, and it will be the poor who can least afford the "genetic upgrades" that may be possible in the second half of the twenty-first century and beyond. It is a logical extrapolation of what happens today, with only the middle and upper classes able to take advantage of technologies such as fertility treatments and cosmetic surgery.

"The fear is often raised that genetic enhancement technologies will be monopolized by the well-to-do at the expense of others, [which] will widen the gap between classes, giving the rich still more advantages," writes Sheila Rothman in the essay, "Redesigning the Self."

"Thus, it will be the 'haves' that will become enhanced, which will provide them with a biological edge in addition to their existing economic edge. These objections, however, no matter how much you may empathize with them, will not slow the drive to enhancement,"

Rothman writes. To wait until everyone can participate seems fantastical, she adds, because no one has ever presumed to limit other technologies, such as the Internet, until a wide cross-section of people can afford them. It is highly unlikely that will come to pass in genetics.[14]

In 2003, during an address to the Commonwealth Club of San Francisco, James Watson weighed in on the issue. "We should also consider whether we should try and improve human life by adding new genetic material to our germ plasm. I'm in favor of going forward, though most of my fellow scientists say they are against it. I believe they don't want to alarm the public by possibilities, which will never exist," he said, referring to science fiction scenarios involving the creation of a dramatically different and improved human race.

Is there an arrangement that society can arrive at that makes the distribution of both essential (i.e., health-related) and nonessential genetic therapies and enhancements more equitable?

> *Only one precept can prevent the dream of twentieth-century eugenics from finding fulfillment in twenty-first century genetic engineering: No matter how far or how fast the science develops, nothing should be done anywhere by anyone to exclude, infringe, repress, or harm an individual based on his or her genetic makeup.*
>
> Edwin Black, War Against the Weak, (London: Four Walls Eight Windows, 2003), pp. 443–444

Some ethicists have proposed that government should pay for genetic enhancements that are essential to health and demonstrably improve our species, such as germ-line gene therapy for Huntington's disease or other heritable disorders. For "vanity" therapy, however—the desire for blonde hair or, say, above-average height or broad shoulders—the burden should be on the individual.

"The knowledge gained [about the human genome] could cure cancer, prevent heart disease, and feed millions," Washington health policy consultant and ethicist Kathi Hanna writes in the closing essay of *The Genomic Revolution*. "At the same time, its improper use can discriminate, stigmatize, and cheapen life through frivolous enhancement technologies. Because of the promise for great good, we all need to understand more about the science and application of human genomics to ensure that the harms do not materialize."[15]

> *Eternal vigilance is the price of liberty.*
>
> Abolitionist Wendell Phillips in an address before the Massachusetts
> Anti-Slavery Society, 1852

"The increase in social and economic disparities in the United States, based on the conservative 'each person fends for themselves' ideology, has shredded notions of social solidarity," says ethicist Bereano. "It could facilitate the emergence of technologically produced genetic variants as societal *übermenschen* [supermen]. Unfortunately, too many scientists are claiming that such a politically repugnant scenario is 'inevitable.' It's like the misogynist's response to the would-be rape victim, to 'lie back and enjoy it.'

"Those of us who support democracy and equal rights must actively oppose such eugenic experiments with the human genome," Bereano says.

One Genome, So Many Questions

With the astounding pace of genomic discovery, the DNA sciences are moving rapid fire into the future. Yet the questions—such as who has the right to keep and view your genetic profile, and how will it be used by employers, insurance companies, law enforcement, and other organizations?—are a long way from being solved. Though Congress has raised these questions, at this writing, no new legislation was on the books to specifically protect genetic privacy. Existing laws—such as the Americans with Disabilities Act—helped the employees at Burlington Santa Fe settle their case, observers say.

"I like to think that the Center for Responsible Genetics, the American Civil Liberties Union, and others have raised enough concern about issues of genetic discrimination and privacy that the federal and state legislation already in place will tend to block the development of a 'genetic underclass,' or would be sufficiently strengthened if the threats were to increase," says ethicist Bereano.

"But people should not just sit on their butts and expect their interests will be covered," he adds. "People need to be informed."[16]

Moving Forward

Did this book inform you? I hope it has. My goal was to give you a plain English account of the DNA sciences and what they're bringing to the party so that you can make wise decisions—about your well-being and that of your family and society.

> *The knowledge gained [about the human genome] could cure cancer, prevent heart disease, and feed millions. At the same time, its improper use can discriminate, stigmatize, and cheapen life through frivolous enhancement technologies. Because of the promise for great good, we all need to understand more about the science and application of human genomics to ensure that the harms do not materialize.*
>
> *Ethicist Kathi E. Hanna*

It is by no means an easy road ahead. Every day brings more headlines, more vexing terminology, more difficult societal issues. But at this critical juncture—where, for the first time, it is possible for humans to become gods unto themselves—some serious wisdom is required.

Twenty-five centuries ago, Confucius had this to say:

> By three methods we may learn wisdom: First, by reflec-
> tion, which is noblest; second, imitation, which is easiest;
> and third, by experience, which is the bitterest.

Escaping more bitter experience and proceeding along the path of noble reflection is certainly the best we can hope for in this spell-binding DNA revolution.

Introduction

1. Ivan Noble, "'Secret of Life' Discovery Turns 50," *BBC News*, February 27, 2003. http://news.bbc.co.uk/2/hi/science/nature/ 2804545.stm.

2. "What They Said: the Genome in Quotes," *BBC News*, June 26, 2000. http://news.bbc.co.uk/1/hi/sci/tech/807126.stm. This article contains a collection of oft-cited quotations from the day the Human Genome Project and Celera Genomics Corp. announced the rough draft of the human genome sequence.

3. Interview with Francis Collins, March 21, 2004.

4. The number of inmates freed from death row when I started this book was 138. It was 143 when I finished it. To get an up-to-date count of death row inmates exonerated by DNA evidence, visit the Web site of the Innocence Project (www.innocenceproject.org).

5. Interview with David Baltimore, October 21, 2003.

Chapter 1

1. Interview with Francis Collins, March 21, 2004.

2. Ann Kellan's interview with Eric Lander, as part of CNN's *Blueprint of the Body* series (June 1, 2000); full text available at http://www.cnn.com/SPECIALS/2000/genome/story/interviews/lander.html. This interview was free-form and wide ranging, covering issues as diverse as gene patenting, the relationship between genes and proteins, so-called junk DNA, and ethics.

3. Interview with David Baltimore, president of the California Institute of Technology (Caltech), October 21, 2003. Baltimore was only thirty-seven when he received the Nobel Prize in 1975 for his work in virology. He is widely considered to be one of the most influential biologists of our time.

4. Stephen Frazier's interview with Francis Collins on CNN's *NewsStand*, May 1999, as part of CNN's extensive *Blueprint of the Body* series; full text available at http://www.cnn.com/SPECIALS/2000/genome/story/interviews/collins.html. In it, Collins also told Frazier that he considered the sequencing of the human genome "more significant than splitting the atom or going to the moon." But he qualified his remarks, saying that people shouldn't confuse the mapping of a gene or the sequencing of the genome with the creation of real disease cures and therapies.

5. Interview with David Baltimore, October 21, 2003.

6. Interview with Leroy Hood, October 21, 2003.

Chapter 2

1. Interview with David Galas, November 24, 2003.

2. Ibid.

3. Nearly a century later, Paul Weindling analyzed Oscar Hertwig's work in his 1991 book, *Darwinism and Social Darwinism in Imperial Germany*, stating that Hertwig's image of the sun rising "conveyed the discovery of the moment at which a new life was formed." The metaphor aptly describes, he said, what Hertwig must have intuitively understood: That the sum (life) was greater than the union of the two sex cells.

4. Robin Marantz Henig, *The Monk in the Garden: The Lost and Found Genius of Gregor Mendel, The Father of Genetics* (New York: Mariner Books, 2000), p. 170. This book tells the interesting story you never learned in high school—how Mendel was largely ignored when he released the data from his groundbreaking experiments with peas, and what happened along the road from obscurity to his eventual recognition as the father of genetics.

5. Interview with James Watson, October 12, 2003.

6. J. D. Watson, *The Double Helix* (New York: Atheneum, 1968), pp. 21, 35. Watson has claimed that his goal with this book was to record, as soon after the events as possible, the inside story of the discovery of DNA's structure.

He records the personality conflicts as well as the scientific details, and he pulls no punches.

7. Watson, *The Double Helix*, p. 124.

8. Interview with James D. Watson, October 13, 2003.

9. Francis Crick, *What Mad Pursuit* (New York: Basic Books, 1988), p. 71.

10. James D. Watson, address to the Commonwealth Club of San Francisco, October 7, 2003.

11. Matt Ridley, *Genome: The Autobiography of a Species in 23 Chapters* (New York: HarperCollins, 1999), p. 24.

12. J. D. Watson and F. H. C. Crick, "A Structure for Deoxyribose Nucleic Acid," *Nature* 171 (April 25, 1953), pp. 737–738.

13. Francis Crick, *What Mad Pursuit* (New York: Basic Books, 1988), pp. 62–63. Crick's memoir is a breezy read, and because it doesn't stop at the discovery of the double helix, it doesn't overlap too much with Watson's *The Double Helix*.

14. Richard Dawkins, *River Out of Eden* (New York: Basic Books, 1995), pp. 16–17 I highly recommend this general audience book, which is a witty and engaging explainer of evolution theory and why all of life "almost certainly" derived from a single ancestor.

15. M. Avins, "On the Trail of a Killer, They Discovered Hope," *Los Angeles Times*, November 14, 1999, p. E1.

16. Ridley, *Genome: The Autobiography of a Species in 23 Chapters*, p. 58.

17. Nancy Wexler, "Clairvoyance and Caution: Repercussions from the Human Genome Project," *The Code of Codes*, eds. D. Kevles and L. Hood (Cambridge, MA: Harvard University Press), pp. 211–243.

18. Ibid.

19. Ridley, *Genome: The Autobiography of a Species in 23 Chapters*.

20. Avins, "On the Trail of a Killer, They Discovered Hope."

21. L. M. Smith, et al., "Fluorescence Detection in Automated DNA Sequencing Analysis," *Nature* 321 (1986), pp. 674–678. This is the original paper detailing Leroy Hood's invention of the automatic sequencer.

22. L. Hood, "After the Genome, Where Should We Go?" *The Genome Revolution* (Washington D.C.: Joseph Henry Press, 2002), p. 64.

23. Ridley, *Genome*, p. 57.

24. Leslie Roberts, "Genome Patent Fight Erupts," *Science* (October 11, 1991), p. 184.

25. Interview with Leroy Hood, October 21, 2003.

26. Interview with David Baltimore, October 21, 2003.

27. "Human Genome Project Complete," *The Scientist* (April 15, 2003); available at http://www.biomedcentral.com/news/20030415/03/.

28. Ridley, *Genome*, p. 24.

29. C. Ezzel, "The Business of the Human Genome," *Scientific American* (July 2000), pp. 48–49.

30. Kevin Davies, *Cracking the Genome* (New York: Simon & Schuster, 2001), p. 37.

31. Ridley, *Genome*, p. 13; also Joseph Campbell, *Grammatical Man: Information, Entropy, Language and Life* (London: Allen Lane, 1983).

Chapter 3

1. BBC interview with Craig Venter, June 2000.

2. Stephen Frazier's interview with Craig Venter on CNN's *Newsstand* (June 3, 1999); available at http://www.cnn.com/SPECIALS/2000/genome/story/interviews/venter.html.

3. Harold Freeman's prologue to the Institutional Review Board for the Protection of Human Subjects, in "Race Not Seen as Factor in Variation of Genetic Code," *San Jose Mercury News*, February 20, 2001. Freeman went on to say, "One of the most astonishing features of the contemporary discussion on race is the fact that anthropology, the science that deals with human biological and cultural variation, has managed to be marginalized . . . [R]egardless of reason, it is clear that there is no consensus and great confusion exists in the discipline with regards to race."

4. "Genome 'Treasure Trove,'" *BBC News*, February 11, 2001. http://news.bbc.co.uk/2/hi/science/nature/1164839.stm.

5. Douglas C. Wallace, "Using Maternal and Paternal Genes to Unlock Human History," *The Genomic Revolution*, eds. Rob DeSalle and Michael Yudell (Washington D.C.: Joseph Henry Press, 2002), p. 131.

6. Interview with David Baltimore, October 21, 2003.

7. John Cook, "Junk May Hold the Key to Genome Puzzle," *Seattle Post-Intelligencer*, October 10, 2003, p. D-1.

8. *The Cambridge Encyclopedia of Human Evolution* (Cambridge, England: Cambridge University Press, 1992).

9. "Cracking the Code of Life," *NOVA*, PBS (airdate: April 17, 2001). All the transcripts of this two-hour special, hosted by ABC News correspondent Robert Krulwich, are available online at http://www.pbs.org/wgbh/nova/transcripts/2809genome.html.

10. "Sequencing Life," *The NewsHour with Jim Lehrer*, PBS transcript (February 12, 2001).

11. Ibid.

12. "Cracking the Code of Life," *NOVA*, PBS.

Chapter 4

1. James Dao, "In Same Case, DNA Clears Convict and Finds Suspect," *The New York Times*, September 6, 2003, p. A-7.

2. The 1998 Australia Prize, press materials.

3. Kary Mullis, in *Nobel Lectures, Chemistry* 1991–1995, ed. B.G. Malmstrom (Singapore: World Scientific Publishing Co., 1997).

4. Interview with James Watson, October 12, 2003.

5. Harry Weinstein, "DNA Frees Man Jailed for 22 Years," *Los Angeles Times*, September 20, 2003, Section A. In addition to profiling Willis, the article goes on to highlight the importance of keeping DNA evidence, as opposed to destroying it at a set date. Without the DNA evidence, Willis would still be imprisoned.

6. *The Human Genome*, C. Dennis and R. Gallagher, eds. (Hampshire: Palgrave, 2001), p. 54.

7. Interview with Peter Neufeld, February 25, 2004.

8. Interview with Philip Bereano, January 26, 2004.

9. Interview with Benjamin Keehn, transcript from *The NewsHour with Jim Lehrer* (July 10, 1998). In the same program, Dr. Paul Ferrara, of the Virginia Division of Forensic Science, said, "If you are guilty of a crime, DNA is probably your worst enemy. On the other hand, if you're innocent, it's your greatest friend."

10. Interview with Philip Bereano, January 26, 2004.

11. "DNA: Is the 9th Circuit Wrong?" *National Review Online*, October 14, 2003; available at http://www.nationalreview.com/debates/dna 200310141029.asp.

12. Interview with Troy Duster, January 25, 2004.

13. "Who Shot Billy the Kid?" Voice of America, August 23, 2003.

14. Interview with Svante Paabo, February 18, 2004.

15. Interview with Douglas Wallace, March 2004.

Chapter 5

1. "Life Blood," *ABC News 20/20* transcript (airdate: February 23, 2001).

2. Ibid.

3. Interview with Mark Hughes, November 3, 2003.

4. Ibid.

5. J. D. Watson, *A Passion for DNA* (Cold Spring Harbor, NY: Cold Spring Harbor Laboratory Press, 2000), p. 208.

6. Interview with Mark Hughes, November 3, 2003.

7. Sergio Pistoi, "Facing Your Genetic Destiny," *Scientific American.com*, February 18, 2002; available at www.sciam.com/article.cfm?articleID =00016A09-BESF-1CDA-B4A8809EC588EEDF.

8. Francis Collins, *The Charlie Rose Show*, PBS Television (June 20, 2000).

9. Interview with Leroy Hood, October 21, 2003.

10. Pistoi, "Facing Your Genetic Destiny."

11. Nancy Wexler, "Clairvoyance and Caution: Repercussions from the

Human Genome Project," *The Code of Codes*, D. Kevles and L. Hood, eds. (Cambridge, MA: Harvard University Press, 1992), p. 238.

12. Shaoni Bhattacharya, "Gene to Halt Ovarian Cancer Found," *Nature Genetics;* available at New Scientist online news service (June 23, 2003), http://www.newscientist.com/news/news.jsp?id=ns 99993859.

13. Wexler, "Clairvoyance and Caution: Repercussions from the Human Genome Project," p. 226.

14. "Gene Raises Heart Attack Risk," *BBC News World Edition*, January 1, 2004.

15. James Dwyer, et al., "Arachidonate 5- lipooxygenase promoter genotype, dietary arachidonic acid, and atherosclerosis," *The New England Journal of Medicine* 1, 350 (January 1, 2004): 4–7.

16. Interview with Rabbi Joseph Ekstein, *The DNA Files* radio series, SoundVision Productions, 1998.

17. Interview with Ruth Ricker, *The NewsHour with Jim Lehrer,* PBS (airdate: April 3, 1996).

18. Nicholas Parasie, "Bill Aims to Regulate Gene Data," Boston University Washington News Service, October 15, 2003.

19. Ibid.

20. Jim Abrams, "Senate Bill Bans Bias Based on Genetic Information," Associated Press, October 15, 2003.

21. Marcy Darnovsky, "Sex Selection Moves to Consumer Culture," *Genetic Crossroads Newsletter,* No. 33, August 20, 2003; available at http://genetics-and-society.org/newsletter/archive/33.html#II.

22. "Choosing Your Baby's Gender," *CBSnews.com*, November 7, 2002; available at http://www.cbsnews.com/stories/2002/11/06/ earlyshow/contributors/emilysenay/main528404.shtml.

23. Lisa Belkin, "Getting the Girl," *The New York Times Magazine*, July 25, 1999, p. 26.

24. Interview with Lee Silver aired on *The DNA Files* radio series, SoundVision Productions, 1998.

25. Interview with Lee Silver, *Leviathan: Back to the Future*, BBC News (broadcast January 1, 2000); transcript available at http://news.bbc.co.uk/ hi/english/static/special_report/1999/12/99/back_to_the_future/lee_silver.stm.

26. Interview with Mark Hughes, November 3, 2003.

27. deCODE genetics press release, December 11, 2003.

Chapter 6

1. Interview with Steven Austad, November 26, 2003.

2. Ibid.

3. Karen Wright, "Staying Alive," *Discover*, November 2003, p. 64.

4. Interview with Cynthia Kenyon, November 16, 2003.

5. Transcript of Cynthia Kenyon's speech at the "DNA: 50 Years of the Double Helix" conference, Cambridge 2003.

6. Interview with Cynthia Kenyon, November 16, 2003.

7. Ibid.

8. P. E. Slagboom, S. Droog, and D. I. Boomsma, "Genetic determination of telomere size in humans: A twin study of three age groups," *American Journal of Human Genetics* 55: 876–882.

9. "Avian Anti-Aging Secret," *Science*, Vol. 300, No. 5626 (June 13, 2003), p. 1653.

10. Ibid.

11. "Scientists Identify Chromosome Location of Genes Associated with Long Life," *Harvard Gazette*, August 28, 2001; available at http://www.news. harvard.edu/gazette/2001/08.16/chromosomes.html. In a press announcement, study coauthor Thomas Perls said, "This is the first study to use humans to try to find genes that play a role in lifespan . . . [m]any investigators thought longevity was far more complex a trait that wouldn't be influenced by just a few genes."

12. T. Perls and D. Terry, "Understanding the Determinants of Exceptional Longevity," in *Annals of Internal Medicine* (September 2, 2003): 445–449.

13. Interview with Aubrey de Grey, December 3, 2003.

14. Jonathan Leake, "Science Gets Serious About Elixir of Life," London's *The Sunday Times*, August 31, 2003, p. 10.

15. Interview with Aubrey de Grey, December 3, 2003.

16. Ibid.

17. Biogerontologist Richard Miller, interview with the American Association for the Advancement of Science, Washington D.C, May 28, 2003.

18. Interview with Richard Miller, January 26, 2004.

19. Ibid.

20. Interview with Steven Austad, January 28, 2004.

21. Biodemographer James W. Vaupel, in a speech at the University of Michigan on December 12, 1999.

22. Matt Ridley, *Genome* (New York: Harper Collins Publishers, 1999), p. 204.

Chapter 7

1. Steve Bunk, "Into the Future," Special Supplement on Cancer, *The Scientist*, September 22, 2003.

2. Jo Revill, "The Cancer Revolution," *The Observer*, March 9, 2003, p. 18.

3. A. Yamamura, *Impact of Genomics on Cancer Diagnosis, Therapeutics, and Pharmacogenics* (Cambridge Healthtech Institute, February 2000).

4. "Careful Planning Guides Center's Strategies to End Disparities by 2015," *National Cancer Institute Newsletter* Vol. 1, Issue 2 (Fall 2003); available at http://crchd.nci.nih.gov/news/newsletters/vol1_issue2/story3.htm.

5. Catherine Arnst, "Cancer: The Hope, the Hype, the Reality," *BusinessWeek*, November 25, 2002, p. 110.

6. Interview with David Galas, November 24, 2003.

7. Project summary of the Cancer Genome Project, The Sanger Institute; available at http://www.sanger.ac.uk/CGP/.

8. Revill, "The Cancer Revolution."

9. American Society of Clinical Oncology press briefing, June 1, 2003.

10. Interview with David Baltimore, October 21, 2003.

11. Jennifer Kahn, "The End of Cancer (As We Know It)," *Wired*, August 2003, p. 108.

12. Interview with Brian Druker, *ABC News.com*, May 10, 2001; available at Healthology Inc. Web site at http://www.healthology. com/focus_article. asp?f=leukemia&b=healthology&c=cml_newdrug.

13. Matt Ridley, *Genome* (New York: HarperCollins Publishers, 1999), p. 190.

14. Interview with Leroy Hood, October 21, 2003.

15. "New Breast Cancer Gene Discovered," *BBC News*, November 26, 2003. http://news.bbc.co.uk/2/hi/health/3234354.stm.

16. University of Cambridge press release, November 26, 2003.

17. Simon J. Boulton, et al., "BRCA1/BARD1 Orthologs Required for DNA Repair in Caenorhabditis Elegans," *Current Biology* 14 (January 6, 2004): 33–39.

18. The Angiogenesis Foundation press conference, September 29, 2003.

19. Interview with Robert Weinberg, October 21, 2003.

20. "Cancer: A Realistic Assessment," *BusinessWeek*.

21. Interview with Howard Chang, January 26, 2004; see also "Wound-Healing Genes Influence Cancer Progression, Say Stanford Researchers," *Medical News Today*, January 13, 2004; available at http://www.medicalnews-stoday.com/index.php?newsid=5310. See also Howard Y. Chang, et al., "Gene expression signature of fibroblast serum response predicts human cancer expression: Similarities between tumors and wounds," *PLoS Biology* Vol. 2, Issue 2 (February 2004).

22. Kahn, "The End of Cancer (As We Know It)."

23. Ibid.

Chapter 8

1. "Scientists Call for Ban on 'Cowboy Cloners,'" Reuters News Service, January 20, 2004.

2. P. Cohen, "Plan to Make Human Cloning Safe Set Out," *New Scientist.com* news service, October 31, 2003. www.newscientist.com/news/news.jsp?id=ns99994334.

3. Tim Friend, "The Real Face of Cloning," *USA Today*, January 17, 2003, p. A.01.

4. A. Coghlan, "Clones Contain Hidden DNA Damage," *New Scientist.com* news service, July 6, 2001. www.newscientist.com/news/news.jsp?id=ns 9999982.

5. Ibid.

6. Friend, "The Real Face of Cloning."

7. "Stem Cell Progress on Parkinson's," *BBC News*, September 15, 2003; available at http://news.bbc.co.uk/2/hi/health/3110364.stm.

8. "U.S. Postpones Global Human Cloning Ban," *NewScientist.com* news service, November 7, 2003. www.newscientist.com/news/news.jsp?id=ns 99994359.

9. J. B. Cibelli, R. P. Lanza, M. D. West, et al., "The First Human Cloned Embryo," *Scientific American*, January 2002, pp. 44–51.

10. Interview with Rohm and Lanza, December 15, 2003.

11. "U.S. Postpones Global Human Cloning Ban," *NewScientist.com*.

Chapter 9

1. Terence Chea, "Gene Therapy's Hot Seat," *Washington Post*, February 20, 2001, p. E1.

2. John C. Fletcher, "Evolution of Ethical Debate About Human Gene Therapy," *Human Gene Therapy*, Vol. 1, No.1 (Spring 1990): 55–65.

3. S. Hart, "New Baldness Gene Found," *ABCNews.com*, January 1998.

4. Interview with Bruce Sullenger, January 15, 2004.

5. P. Jacobs, "Pioneer Genetic Implants Revealed," *Los Angeles Times*, October 8, 1980.

6. Sally Lehrman, "Virus Treatment Questioned After Gene Therapy Death," *Nature*, 401 (October 7, 1999): 517–518.

7. Carol Smith, "Curing Disease from Inside the Cell: 50 Years After DNA Breakthrough, Seattle Is a Leader in Gene Research," *Seattle Post-Intelligencer*, February 28, 2003, p. A-1.

8. Utpal P. Dave, Nancy A. Jenkins, and Neal G. Copeland, "Gene Therapy Insertional Mutagenesis Insights," *Science* (January 16, 2004): 333.

9. "New Clues in 'Bubble Boy' Gene Therapy," *CNN.com* Health, January 15, 2004; available at http://www.cnn.com/2004/HEALTH/01/15/gene.therapy.ap/index.html.

10. Susan Dentzer, "Gene Therapy," *The NewsHour with Jim Lehrer* transcript (airdate: February 2, 2000).

11. Ibid.

12. Larry Thompson, "Human Gene Therapy: Harsh Lessons, High Hopes," *FDA Consumer* magazine (September-October 2000). www.fda.gov/fdac/features/2000/500_gene.html.

13. Chea, "Gene Therapy's Hot Seat."

14. Thompson, "Human Gene Therapy: Harsh Lessons, High Hopes."

15. Vita Foubister, "Genes Go Incognito into Brain," Children's Neurobiological Solutions newsletter, Issue 40 (May 13, 2003).

16. Ibid.

17. Danny Penman, "Subtle Gene Therapy Tackles Blood Disorder," *New Scientist* (October 11, 2002); and M. M. Vacek, et al., "High-level expression of hemoglobin A in human thalassemic erythroid progenitor cells following lentiviral vector delivery of an antisense snRNA," *Blood* (January 1, 2003): 104–111.

18. Ibid.

19. Bob Holmes, "Gene Therapy May Switch Off Huntington's," *New Scientist.com* news service, March 13, 2003. www.newscientist. com/news/news.jsp?id=ns99993493.

20. W. French Anderson, "Gene Therapy: The Best of Times, the Worst of Times," *Science*, 288 (April 28, 2000): 627–629.

21. Robin McKie, "By Human Design: Children of the Revolution," *The Observer*, October 26, 2003, p. 56.

22. S. O. Freytag et al., "Phase I Study of Replication-Competent Adenovirus-Mediated Double-Suicide Gene Therapy in Combination with Conventional-Dose Three-Dimensional Conformal Radiation Therapy for the Treatment of Newly Diagnosed, Intermediate- to High-Risk Prostate Cancer," *Cancer Research*, 63 (November 1, 2003): 7497–7506.

23. McKie, "By Human Design: Children of the Revolution."

24. James Watson, address to the Commonwealth Club of San Francisco, October 9, 2003.

Chapter 10

1. Francis Galton, "Eugenics: Its Definition, Scope, and Aims," *The American Journal of Sociology* Vol. X, No. 1 (July 1904).

2. Edwin Black, *War Against the Weak*. (London: Four Walls Eight Windows, September 2003), p. xvi.

3. Daniel J. Kevles, "Eugenics, the Genome, and Human Rights," *The Genomic Revolution*, eds. Rob DeSalle and Michael Yudell (Washington D.C.: Joseph Henry Press, 2002), pp. 147–154.

4. Kristen Philipkoski, "Blaming the 'Defective' People," *Wired* News, March 26, 2001. www.wired.com/news/technology/0,1282,42567,00.html

5. Steve Irsay, "Cold Hits Versus Civil Liberties: The Looming Debate Over Privacy and DNA Databases," *CourtTV.com*, April 24, 2003; available at http://www.courttv.com/news/forensics/ dna_anniv/databases.html.

6. Interview with Troy Duster, January 25, 2004.

7. Interview with Philip Bereano, January 28, 2004.

8. Written statement of J. Craig Venter, on behalf of the Biotechnology Industry Organization (BIO) before the subcommittee on consumer protection, U.S. House Committee on Energy and Commerce, July 11, 2001.

9. Bob Calandra, "Genetic Testing: Consumers Fear It Will Be Used to Deny Coverage and Raise Premiums," *Risk & Insurance* (April 14, 2003).

10. Interview with Philip Bereano, January 28, 2004.

11. Diane Martindale, "Pink Slip in Your Genes," *Scientific American* (January 2001).

12. Interview with Helen Wallace, January 26, 2004.

13. Ibid.

14. David Rothman and Sheila Rothman, "Redesigning the Self: The Promise and Perils of Genetic Enhancement," *The Genomic Revolution*, eds. Rob DeSalle and Michael Yudell (Washington D.C.: Joseph Henry Press, 2002), pp. 155–164.

15. Kathi E. Hanna, "Summing Up: Finding Our Way Through the Revolution," *The Genomic Revolution*, Rob DeSalle and Michael Yudell, eds. (Washington D.C.: Joseph Henry Press, 2002), pp. 199–208.

16. Interview with Paul Bereano, January 28, 2004.

GLOSSARY

Achondroplasia   A type of inherited dwarfism.

Acquired mutations   Noninherited gene changes that accumulate during a person's lifetime; also called somatic mutations.

Active site   The part of a protein where a chemical reaction occurs, usually by interaction with an enzyme or antibody. This part must stay in a specific three-dimensional shape in order for the protein to be functional. For instance, the active site of an enzyme is the physical point where it binds to a substrate.

Acute myelogenous leukemia (AML)   A malignant bone marrow disease.

Adenine (A)   One of the four bases in DNA, pairing with thymine. (In RNA, it instead pairs with thymine's replacement, uracil.)

Adenosine triphosphate   The energy molecule of cells that is synthesized mainly in mitochondria and chloroplasts; it drives many important cellular reactions. (*See also* ATP.)

Adenovirus   A virus that causes the common cold, respiratory infections, conjunctivitis, and other maladies.

Affected relative pair   Two blood-related individuals, each of whom is affected with the same trait. There are, for instance, affected sibling, cousin, and avuncular pairs. (*See* avuncular relationship.)

Agonists   Small protein molecules that bind to receptor proteins, causing a change in cell activity.

Ala   The abbreviation for alanine, one of the amino acid building blocks for a protein. (*See* amino acid.)

Albino   A pigment-less white phenotype, caused by a mutation in a gene coding for a pigment-synthesizing enzyme.

Allele   A variant form of the same gene. Different forms of a gene cause variations in inherited characteristics such as eye color.

Allele frequency   A measure of how common an allele is in a population; also called gene frequency.

Alzheimer's disease   A neurological disease resulting in progressive dementia and memory loss.

Amino acid   The basic building block of proteins. There are twenty amino acid molecules that combine to form all the proteins present in living things.

Amniocentesis   A technique for testing the genotype of an embryo or fetus in utero.

Amplification of DNA   The production of many DNA copies from just one or a few samples; typically performed using repeated cycles of heating and cooling and exposure to a special thermostable enzyme that is derived from bacteria.

Amyotrophic lateral sclerosis (ALS)   An inherited, fatal degenerative nerve disorder. Also known as Lou Gehrig's disease.

Anemia   A disease resulting in a shortage of red blood cells, which keeps the body from carrying oxygen to the body's tissues and organs. Symptoms include fatigue, rapid heartbeat, shortness of breath, confusion, and fainting.

Angiogenesis   The process the body uses to form and develop new blood vessels.

Animal model   *See* model organisms.

Antagonists   The molecules that bind to a protein's receptor site. Antagonists suppress function of the proteins they bind to. (*See also* agonists.)

Antibiotic   Substances that are able to destroy or inhibit the growth of microorganisms.

Antibody   A Y-shaped protein component of the body's immune system response to a foreign substance (antigen) such as a toxin or bacterium. Each antibody recognizes and binds to a specific antigen.

Antigen   Any foreign substance that, when introduced into the body, causes the immune system to create an antibody.

Anti-oncogene   A gene that prevents malignant tumor growth. When absent by mutation, it results in a malignancy. (An example is the disease retinoblastoma.)

Antisense RNA   An RNA product that regulates genes by base-pairing with a matching RNA pair and thus canceling its action out.

Apoptosis   The cell's self-destruct mechanism, sometimes also referred to as "programmed cell death." The cell dies by self-digesting, disintegrating without rupturing or spilling its contents into surrounding tissue. Without normal cell apoptosis, cells can grow uncontrollably, causing cancer.

Arg   The common abbreviation for arginine, one of the twenty amino acid building blocks of a protein. (*See* amino acid.)

Asexual reproduction   Any creation of offspring by cloning, budding, or other means not involving the combination of genetic material from two individuals.

Asn   The common abbreviation for asparagine, one of the twenty amino acid building blocks of a protein. (*See* amino acid.)

Asp   The common abbreviation for aspartate, one of the twenty amino acid building blocks of a protein. (*See* amino acid.)

Assay   The process of testing a sample of a chemical to see if there is activity against a specific target or cellular response.

Ataxia-telangiectasia   A disease involving loss of muscle control and reddening of the skin caused by radiation damage to DNA.

Atom   The smallest component of an element that still retains all the properties of the element.

ATP   Adenosine triphosphate; the energy molecule of cells, synthesized mainly in mitochondria and chloroplasts; energy from the breakdown of ATP drives many important reactions in the cell.

Autonomous replication sequence (ARS)   A segment of a DNA molecule that is necessary for its replication to begin.

Autosomal dominant   A gene on one of the nonsex chromosomes that is always expressed, even if there is only one copy present. The chance of a person passing the gene along to each child is 50 percent. An example of an autosomal dominant disease is Huntington's chorea.

Autosome   Any chromosome with the exception of the sex-determining chromosomes, X and Y. Human cells have twenty-two pairs of autosomes, also known as the autosomal set.

Avuncular relationship   The genetic relationship between uncles and aunts and their nieces and nephews.

B cells   The cells found in many organs that make antibodies.

Bacterium   A single-celled organism that is the most diverse life-form on the planet. Found throughout nature and in every conceivable habitat, bacteria can be beneficial or harmful. In biotechnology, bacteria take part in a wide range of experiments and processes.

Bacterial artificial chromosome (BAC)   A delivery mechanism, or vector, used to clone DNA fragments. (*See* vector.)

Bacteriophage   A virus that primarily targets and infects bacteria.

Base   Any of four molecular units, known as nucleotides, that are found in DNA. The bases are: adenine (A), cytosine (C), guanine (G), and thymine (T).

Base pairs   Two bases held together by chemical bounds, comprising the "rungs" of the DNA ladder. Adenine is always paired with thymine; cytosine is always paired with guanine. (*See* base.) The human genome has about six billion base pairs.

Base sequence   The order of bases along a DNA strand that determines the structure of proteins.

Base sequence analysis   An automated method of determining a gene's (or genome's) base sequence.

Behavioral genetics   The study of how genes influence behavior.

Biallelic markers   DNA markers that occur in only two forms in a population.

Bioassay   A measure of a drug's effect on animals, tissues, or other organisms and how it compares with a standard preparation.

Biochip   An electronic device (e.g., a semiconductor) with a grid that holds organic molecules.

Bioinformatics   The DNA science of building and deploying tools to help researchers build better experiments. In genome projects, it encompasses methods of faster DNA sequencing and database searching in order to get better protein sequence and structure predictions from DNA data. Bioinformatics also includes computer techniques such as 3D modeling.

Biomarker   Any detectable biological molecule that scientists can consistently associate with a biological state—for instance, a disease.

Biotechnology   The applied science of biological research to drug discovery, medical diagnostics and devices, and techniques for crop improvement and animal health.

Birth defect   A harmful biochemical or physical trait that is present at birth and is often the result of a genetic mutation. (*See also* congenital; mutation.)

BLAST   A computer program designed to identify similar (homologous) genes in different organisms such as the human, the fruit fly, and the roundworm.

Bone marrow transplantation   The transfer of bone marrow—the blood-cell-producing tissue found in bone cavities—from one patient to another.

BRCA1, 2   A gene that normally restrains cell growth.

BRCA1, 2 breast cancer susceptibility genes   The mutated version of BRCA1 that predisposes a person toward developing breast cancer.

Cancer   A family of diseases in which abnormal cells divide and grow out of control. Some cancers spread from their original site to other parts of the body. Left unchecked, most cancers are fatal. (*See also* hereditary cancer; sporadic cancer.)

Candidate gene   A gene located in a chromosome region that researchers suspect could be involved in a certain disease. (*See* positional cloning.)

Carcinogen   A substance or environmental factor that causes changes in a cell's DNA, resulting in cancer.

Carrier   Someone who has a recessive mutated gene as well as its normal version (allele) in a pair. Carriers don't usually suffer from the disease they carry, but they can pass it on to their children.

Carrier testing   The process of identifying individuals who may be carriers of recessive gene disorders. Carrier testing targets healthy people—prospective parents, for instance—who don't have symptoms of a disease but need to know if they have recessive genes that put future children at risk.

Cell   The basic unit of any living thing; it is a tiny, watery compartment containing a nucleus and surrounded by a fatty membrane.

Chemical base   Adenosine, thymine, cytosine, and guanine—A, T, C, and G—the building blocks of DNA. (*See also* base; base pairs.)

Chemotherapy   The use of toxic chemicals to poison cancer cells and treat the disease. Chemotherapy targets rapidly replicating cells that typify cancer.

Chromosomal deletion   The loss of a part of the DNA on a chromosome.

Chromosomal inversion   Segments on a chromosome that have been turned 180 degrees. The gene sequence for the segment is reversed and compared to the gene sequences on the rest of the chromosome.

Chromosome   The structure found in a cell's nucleus on which the genes are located. Chromosomes come in pairs. A normal cell contains twenty-three pairs—twenty-two pairs of autosomes and one set of sex chromosomes (two Xs, or an X and a Y).

Chromosome region p   A term referring to the short arm of a chromosome.

Clade   A group of individuals, characteristics, or DNA sequences that are obviously related to one another.

Clinical trial   A scientific study allowing researchers to try FDA-unapproved drugs and therapies on animals and humans. A phase I clinical trial studies the effect of a new drug on humans. A phase II trial studies drug safety, efficacy, and side effects. Phase III compares the new drug to existing therapies for the same disease or condition. Phase IV trials test an already-approved drug for new indications or on new patient populations.

Clone   A group of identical genes, cells, or organisms derived from a single sample.

Cloning   The process of making genetically identical copies. The process of cloning identical sequences of base pairs, which was used by researchers in the Human Genome Project, is referred to as cloning DNA. The resulting cloned (i.e., copied) collections of DNA molecules are called clone libraries. A second type of cloning exploits the natural process of cell division to make many copies of an entire cell. The genetic makeup of these cloned cells, called a cell line, is identical to the original cell. A third kind of cloning produces complete, genetically identical animals to the DNA donor—such as that famous Scottish sheep, Dolly.

Codon   A section of DNA, three bases in length, that functions as the "words" of DNA. Each three-base codon is a code for an amino acid. Twenty amino acids form all the proteins in the human body.

Comparative genomics   The study of human genetics by comparing model organisms, such as the fruit fly, the mustard plant, the E. coli bacterium, the mouse, and the human.

Complementary DNA (cDNA)   Single-stranded DNA that is synthesized by using a messenger RNA sample.

Complementary sequence   A sequence of DNA bases that, because of the rules of pairing, automatically forms a double-stranded structure with a second sequence of DNA bases. The pairing rules are simple: A always pairs with T; C always with G. For example, the complementary sequence of the strand GCTA would be CGAT.

Complete response   The complete disappearance of a tumor resulting from a treatment.

Complex trait   A trait or disorder that is genetic, yet does not follow strict Mendelian inheritance. Complex traits likely involve the interaction of two or more genes, or the interaction of genes and the environment. (*See* Mendelian inheritance.)

Congenital   A trait that is present at birth, regardless of whether it is caused by genetics or environment. (*See also* birth defect.)

Conserved sequence   A DNA base sequence that has remained practically unchanged throughout evolution. This term can also apply to conserved amino acid sequences.

Constitutive ablation   Any gene expression that results in the death of a cell.

Crossing over   Also known as recombination, this phenomenon sometimes occurs during the formation of egg and sperm cells, when a pair of chromosomes (one each from the mother and father) break and trade segments with one another.

Cystic fibrosis (CF)   An inherited disease that causes thick mucus to clog the lungs and block the pancreatic ducts.

Cytogenetics   The study of how chromosomes physically appear. (*See* karyotype.)

Cytokine   Any chemical that triggers cell division.

Cytological band   An area of the chromosome that, when dyed, stains differently from areas around it.

Cytoplasm   The cellular substance outside the nucleus; the cell's organelles are suspended in it.

Cytoplasmic trait   A characteristic where genes are located outside the nucleus—in chloroplasts or the mitochondria. In humans, results in the mitochondrial genes coming only from the mother.

Cytosine (C)   One of the four bases of DNA, pairing with guanine (G).

Deletion   The loss of part of the DNA from a chromosome, potentially leading to abnormalities or disease. (*See also* chromosome; mutation.)

Deletion map   A description of a specific chromosome that uses defined deletions as markers to indicated specific areas.

Dementia   A severe impairment of mental function.

Deoxyribonucleic acid   *See* DNA.

Deoxyribose   The type of sugar that comprises part of DNA.

Diabetes   A disorder in which the body is unable to maintain correct glucose levels in the blood.

Diploid   The full set of all the paired chromosomes in an organism's genetic material. In animals, all cells except the sex cells (gametes) have a paired, or diploid, set of chromosomes. The diploid human genome includes forty-six chromosomes grouped as twenty-three pairs. (*See* haploid.)

Disease-associated genes   Certain DNA sequences associated with the presence or high likelihood of disease.

DNA   Deoxyribonucleic acid. DNA is the substance of heredity present in the nucleus of almost all cells in an organism. This large molecule, shaped like a double helix, carries genetic information that cells need to replicate and to produce proteins that govern all life processes.

DNA bank   A commercial service that stores DNA from blood samples or other tissue samples.

DNA ligase   The enzyme responsible for "gluing," or joining, ends of DNA segments (which are usually double-stranded) to form a DNA chain.

DNA marker   Unique DNA sequences used by researchers to characterize or keep track of a gene, chromosome, or DNA lineage.

DNA probe   *See* probe.

DNA repair genes   Genes coding for proteins that correct errors in DNA sequencing. If altered, they permit mutations to pile up in DNA.

DNA replication   The use of existing DNA as a template to create new DNA strands. This process works because of the way the bases match in DNA—A always with T and C always with G. Therefore, a strand that reads ACTG will always bind with the matching strand TGAC.

DNA sequencing   Determining the exact order of the base pairs in a segment of DNA, usually by automated means.

Domain   A portion of a protein with its own function. The combination of several domains in a protein determines its overall function.

Dominant allele   A mutation that is expressed even if its counterpart on the other chromosome is normal. A single mutation of a dominant allele is responsible for such autosomal dominant disorders as Huntington's disease. (*See also* recessive allele.)

Double helix   A shape reminiscent of a twisted ladder or staircase. This is the shape of the DNA molecule with its two linear strands as sides of the ladder and the complementary bases bonding together as rungs. This shape allows cells to pack DNA tightly in the tiny area of a cell's nucleus.

Draft sequence   The DNA sequence generated by Celera Genomics and the Human Genome Project in June 2000. It included the sequences and chromosomal locations of about 95 percent of human genes. In April 2003, the corrected, final sequence was published.

Efficacy   How well a drug treats an illness.

Electroporation   A process that uses high-voltage currents to make cell membranes permeable enough to allow the introduction of new DNA. It is a commonly used technique in recombinant DNA technology.

Embryo   An organism in its early stage of development, after conception but before the formation of major organs. (*See* fetus.)

Embryonic stem (ES) cell   An early version of a cell that is able to transform into virtually any type of cell; it can replicate almost indefinitely and serves as a continuous source of new cells.

Endonuclease   *See* restriction enzyme.

Enzyme   Any protein that helps facilitate a specific chemical reaction; it acts as a catalyst but in no way affects the direction or type of reaction.

Epistasis   One gene interfering with or overriding the expression of another.

Escherichia coli   Also known as E. coli, this common bacterium is frequently studied by geneticists because of its small genome size, rapid growth rate, and ease of maintenance in the laboratory.

Eugenics   The study of improving a species by selective breeding. Its most negative association is with Nazi Germany, which applied the concept to the sterilization of hundreds of thousands of people and eventually the murder of millions of Jews, gypsies, and "social undesirables."

Eukaryote   Any cell or organism that includes a cell nucleus. The term describes all organisms on earth, with the exception of viruses, bacteria, and blue-green algae. (*See* prokaryote.)

Evolutionarily conserved   *See* conserved sequence.

Exogenous DNA   Any DNA in an organism that originally came from the DNA of another organism.

Exon   The portion of the DNA in a gene that codes for protein, as opposed to regulating other genes or performing unknown functions. (*See also* intron.)

Expressed gene   *See* gene expression.

Expressed sequence tag (EST)   A short strand of DNA that can act as an identifier of a gene. ESTs are commonly used for locating and mapping genes. (*See also* sequence tagged site.)

Ex vivo gene transfer   The transfer of genetic material to cells located outside their original location. The cells with their new genetic material are then transplanted back into the cells' original location; this is also commonly referred to as the indirect method of gene transfer.

Familial cancer   Cancer or a predisposition toward cancer that runs in families.

Fetus   An organism in early development, after its major organs are formed.

Food and Drug Administration (FDA)   The U.S. federal agency that's responsible for regulating many health, drug, and food products. The FDA, for instance, regulates gene therapy experiments.

Founder effect   A change in the gene pool of a colonizing population reflecting a limited number of individuals in its parent population.

Founding lineage   The DNA present in the original founders of a population.

Functional gene tests   Biochemical assays for a specific protein, indicating which genes are present and active.

Functional genomics   The study of what genes do, in order to determine the roles genes play in disease and other biological processes.

Gamete   Mature male or female reproductive cells (sperm or ovum). Each has a haploid set (no pairs) of chromosomes, twenty-three in all humans.

GC-rich area   A long stretch of DNA with many repeated Gs (guanine) and Cs (cytosines); often indicates a gene-rich region.

Gene A fundamental unit of inheritance. Genes are made of DNA and lie on chromosomes. There are an estimated 30,000 genes in the human genome directing the body's manufacture of proteins.

Genealogy An account of a person or a family's descent through one or more ancestral lines.

Gene amplification The repeated copying of a piece of DNA; it is also a characteristic of tumor cells. (*See also* gene; oncogene.)

Gene chip technology The development of complementary DNA (cDNA) microarrays using a large number of genes that researchers can use to monitor and measure changes in gene expression.

Gene deletion The loss or absence of a gene.

Gene expression The process through which a gene's coded information is translated into the structures present and operating in the cell (either proteins or RNA molecules).

Gene family A group of closely related genes that make similar proteins.

Gene mapping The effort of determining the relative positions of genes on a chromosome and the distance between them.

Gene markers Landmarks that researchers use to identify a target gene, either distinctive segments of DNA or detectable traits of that gene.

Gene pool All variations of genes in a species. (*See also* allele; polymorphism.)

Gene prediction The practice of using a computer program to predict possible genes, based on how well a stretch of DNA sequence matches up to known gene sequences.

Gene product The biochemical material that results from the expression of a gene, always either RNA or protein. The amount of gene product directly correlates to how active a gene is. Abnormal amounts can be indicative of disease-causing gene mutations.

Gene testing Examining a blood or body fluid sample for biochemical, chromosomal, or genetic markers that indicate the risk or the presence or absence of genetic disease.

Gene therapy The treatment of disease by replacing, manipulating, turning off (i.e., knocking out), or supplementing mutated, nonfunctional genes.

Gene transfer   The incorporation of new DNA into an organism's cells, usually by a modified virus or other vector. Typically, gene transfer is used in gene therapy. (*See also* gene therapy; mutation; vector.)

Genetic counseling   A key part of gene testing that provides patients and families with education and information about genetic conditions.

Genetic discrimination   Prejudice against those who have or are likely to develop an inherited disorder in their lifetimes.

Genetic distance   A measure of how related two or more populations are based on how frequently they share genes.

Genetic drift   Changes in a small population's gene pool due to chance.

Genetic engineering   The practice of changing the genetic material of cells or organisms to help them to make new substances or perform different functions.

Genetic engineering technology   *See* recombinant DNA technology.

Genetic fingerprint   *See* genetic profiling.

Genetic illness   A hereditary disorder, sickness, or physical disability.

Genetic informatics   *See* bioinformatics.

Genetic linkage maps   DNA maps that denote relative chromosomal locations of gene markers (either genes for known traits or distinctive DNA sequences), based on how frequently they are inherited together.

Genetic material   *See* genome

Genetic polymorphism   Differences in DNA sequences among individuals or populations (e.g., the gene for blue eyes versus brown eyes).

Genetic predisposition   Susceptibility to a genetic disease. For instance, a test that reveals a mutation on the BRCA2 gene reveals that a person is more susceptible to getting breast cancer, but that susceptibility may not result in an actual disease.

Genetic profiling   Invented in 1985, this technique for comparing DNA samples is also known as DNA fingerprinting. It is commonly used by police agencies and governments to determine the probability of a blood or tissue sample coming from a certain source.

Genetic screening   The testing of a group of people to find those at high risk of having or passing on a certain genetic disorder.

Genetic testing   The testing of an individual's genetic material to discover predisposition for particular genetic disorders or to confirm diagnosis of a genetic disease.

Genetics   The study of heredity, or how parents transmit qualities and traits to their offspring.

Genome   All the genetic material in a particular organism's chromosomes.

Genomics   The study of genes and their function.

Genotype   The actual genes carried by an individual. Genotype is distinct from phenotype, the individual's physical characteristics as manifested from the genes.

Germ cells   The sperm and egg cells; the reproductive cells of the body. Germ cells are haploid and have only one set of chromosomes (twenty-three in all), while all other cells have two copies (forty-six in all).

Germ line   The linear continuation of a set of genetic information from one generation to the next. (*See* inherit.)

Germ-line gene therapy   The highly controversial and as yet experimental process of inserting genes into the sex cells (sperm or ovum) in order to cause genetic changes to offspring in perpetuity. Theoretically, the process could be used to halt familial inheritance of genetic diseases. (*See* somatic cell gene therapy.)

Germ-line mutation   Gene mutations that are passed along by heredity to offspring.

Guanine (G)   One of the four bases of DNA, pairing with cytosine (C). (*See also* base pairs; nucleotide.)

Haploid   A single set of unpaired chromosomes present in the egg and sperm cells of animals and in the egg and pollen cells of plants. This enables reproduction, where the offspring gets twenty-three chromosomes from the mother and another twenty-three from the father. Nonsex cells have a diploid set—that is, forty-six chromosomes in twenty-three pairs.

Haplotype   A method of describing a collective genotype of a number of closely linked genes on a chromosome.

Hemizygous   Having only one copy of a particular gene. For example, in humans, males are hemizygous for genes found on the Y chromosome.

Hemophilia A A disease that affects the process of blood clotting. Patients with hemophilia A are prone to spontaneous, uncontrolled internal bleeding that can lead to restricted mobility, pain, and even death. It is caused by the deficiency or absence of a protein called factor VIII, which is involved in the blood coagulation pathway.

Hereditary cancer A cancer-causing gene mutation that runs in the family. (*See also* sporadic cancer.)

Heredity As first defined by Gregor Mendel, the relationship between successive generations. More specifically, the transmission of characteristics from parents to offspring via the chromosomes that carry DNA.

Heterozygote An organism that has two different versions of an allele, for instance, one for blue and the other for brown eyes.

Highly conserved sequence A DNA sequence that is similar across several different types of organisms. (*See* gene; mutation.)

High-throughput sequencing A rapid way of determining the order of the bases in DNA. (*See also* sequencing.)

Homeobox A short sequence of bases where the sequence is practically identical in all the genes that contain it; the homeobox (or hox) genes seem to determine the positions of body segments in higher organisms. Homeoboxes have been found in many organisms, from fruit flies to human beings.

Homolog A member of a chromosome pair in a diploid organism, or a gene that has the same origin and functions in two or more species.

Homologous chromosome Chromosomes containing the same linear gene sequences as one another, each derived from a parent.

Homologous recombination Swapping of DNA fragments between paired chromosomes.

Homology Similarity in DNA or protein sequences between individuals within the same species or of different species.

Homozygote An organism that has two identical alleles of a gene—for instance, two alleles for blue eyes. (*See also* heterozygote.)

Hormone Chemicals produced by glands in the body that circulate in the bloodstream, often exerting control over various other parts of the body.

Hox genes   *See* homeobox.

Human artificial chromosome (HAC)   A vector used to hold large DNA fragments for gene therapy and other uses. (*See* vector.)

Human genome   The full collection of genes needed to produce all the proteins that produce a human being.

Human Genome Project   An international research effort that was aimed at identifying and ordering every base in the human genome. The finished version of the genome was delivered in April 2003.

Huntington's disease   An adult-onset disease leading to progressive mental and physical deterioration; it is caused by an inherited dominant gene mutation of three bases—C, A, and G—repeated excessively.

Hybrid   The offspring of genetically different parents. (*See* heterozygote.)

Hybridization   The process of joining two complementary strands of DNA, or one each of DNA and RNA, to form a double-stranded molecule.

Identical twin   A twin produced by the division of a single zygote. Each twin has an identical genotype to the other.

Immunotherapy   A method of using a patient's own immune system to treat disease. Vaccines are an example.

Imprinting   A biochemical phenomenon that determines (for some genes) which one of a pair of alleles—the mother's or the father's—will be active in that individual.

Informatics   *See* bioinformatics.

Inherit   A genetic term referring to how offspring get DNA from their parents.

Insertion   Also referred to as an insertional mutation, this occurs when a piece of DNA gets incorporated into a working gene and disrupts its normal function. (*See* gene; mutation.)

Intellectual property rights   Patents, copyrights, and trademarks. (*See* patent.)

Interferon   A class of small proteins possessing potent antiviral effects.

Interleukin-2 (IL-2)   A protein in the human body that's responsible for stimulating the immune system to produce white blood cells.

Intravenous (IV)   Injection of a substance into the bloodstream through a vein.

Intron   A DNA sequence interrupting the protein-coding part of a gene; RNA transcribes it, but the intron is cut out of the message before it gets translated into a protein. (*See also* junk DNA; exon.)

In vitro   An experiment or procedure performed outside of a living organism, such as in a test tube or a petri dish in a laboratory.

In vivo   An experiment or procedure performed inside a living organism.

IQ   Short for intelligence quotient, an index that has been correlated with intelligence.

IRB (Institutional Review Board)   An independent group overseeing clinical trials that typically includes doctors, nurses, social workers, ethicists, and patient advocates. IRBs are in charge of overseeing all gene therapy trials in the United States.

IVF   Short for in vitro fertilization, a method of fertilizing an egg with sperm in a laboratory environment outside the body.

Junk DNA   Long stretches of DNA that do not code information for genes. Most of the genome, in fact, consists of so-called junk DNA, although it probably has regulatory and other functions. Also called non-coding DNA.

Karyotype   An image revealing all of an individual's chromosomes. Researchers use karyotypes to correlate gross chromosomal abnormalities with specific diseases—for instance, to detect an extra or missing chromosome.

Kilobase (kb)   A unit of length for DNA sequences equal to 1,000 nucleotides or base pairs.

Knockout   The purposeful deactivation of specific genes, used in laboratory animals to study gene function. (*See* model organisms.)

Laws of inheritance   The laws formulated by Gregor Mendel in 1860. The law of segregation states that each hereditary trait is controlled by two factors (alleles of a gene) that separate and individually pass into ova and sperm. The law of independent assortment states that pairs of alleles separate independently of each other when reproductive cells are formed.

Leukemia   A type of cancer that begins in the bone marrow's developing blood cells.

Library   A collection of cloned sequences of DNA.

Li-Fraumeni syndrome   Caused by a mutation in the p53 tumor-suppressor gene, this disorder leads to a predisposition to multiple cancers.

Linkage   The proximity of two or more markers (e.g., genes) on a chromosome. The closer the markers, the greater the chances are that offspring will inherit them together. This is why linked genes are an exception to Mendel's laws of inheritance. (*See* laws of inheritance.)

Linkage analysis   A method of finding disease-causing genes by tracing patterns of heredity in large, high-risk families and attempting to spot traits that are coinherited with the disorder.

Linkage map   A map of where genes are located on a chromosome, determined by how often genes are inherited together.

Liposome   An artificial bubble made of fatty molecules that can contain substances, including drugs or DNA sequences, designed to be absorbed by specific cells.

Locus   The physical position on a chromosome of a gene or other marker. The plural of locus is loci. (*See* gene expression.)

Lymphocyte   A small white blood cell that is critical in defending the body against disease. There are two main types of lymphocytes: B cells, which make antibodies to attack toxins and bacteria, and T cells, which attack cells in the body if they have become infected by viruses or have turned cancerous.

Malignant   Cancerous.

Maternal inheritance   DNA that is inherited solely from the mother—for instance, the DNA located inside the mitochondria.

Megabase (Mb)   Unit of measurement for DNA fragments equal to 1 million nucleotides or base pairs.

Melanoma   A cancer that begins in skin cells (called melanocytes) and spreads to internal organs.

Mendelian inheritance   Named for Gregor Mendel, it is a method in which genetic traits are passed from parents to offspring. Mendel was the first to study and recognize the existence of genes. (*See also* autosomal dominant; recessive allele; sex-linked.)

Messenger RNA   *See* mRNA.

Metabolism   The biochemical processes that keep a living cell or organism going.

Metastases   Cancer cells that have detached from the original cancer and spread (metastasized) to other locations in the body, forming new tumors in those places.

Microarray   A chip with tiny chambers to store fragments of DNA, antibodies, or proteins, allowing researchers to perform chemical reactions on many samples at a time.

Mitochondria   The structure within every cell that is the site of energy production. Mitochondria evolved from bacteria and contain their own DNA—distinct from the DNA on the twenty-three pairs of chromosomes inside the cell nucleus—that is only inherited from the mother. Mitochondrial DNA (also referred to as mtDNA) is frequently used to determine the maternal lineage of a person's DNA.

Model organisms   A lab animal or other organism useful for research. Typical model organisms in DNA research include fruit flies, roundworms, mustard plants, and mice.

Molecule   A group of atoms that is physically arranged to interact in a particular way; one molecule of any substance is the smallest physical unit of that particular substance.

Monoclonal antibody   A protein that can be manufactured to specifically bind to any single substance in the body, either to register its presence or to deactivate it.

Monogenic disorder   A disorder caused by the mutation of a single gene. (*See* mutation; polygenic disorder.)

mRNA   Messenger RNA, or a single-stranded molecule of ribonucleic acid that translates information from DNA to the protein-assembling parts of the cell.

mtDNA   *See* mitochondria.

Multifactorial or multigenic disorder   A disorder resulting from mutations on several genes. (*See* polygenic disorder.)

Multiple sclerosis (MS)   A disease in which the body's immune cells attack the insular material surrounding nerve fibers in the spinal cord and brain. MS leads to recurrent muscle weakness, loss of muscle control, and often, eventual paralysis.

Mutagen   A toxin, environmental factor, or other substance that increases the rate of mutation.

Mutant   A cell or organism that manifests new characteristics due to a change in its DNA.

Mutation   Any change in the number or arrangement of a gene or its molecular sequence. A mutation is an inheritable change in the genetic code.

Neo-Darwinism   A synthesis of Mendel's laws of inheritance and Darwin's theory of natural selection.

Neurodegenerative   A term describing diseases such as Parkinson's and Alzheimer's, in which parts of the nervous system deteriorate.

Newborn screening   The examination of blood samples from a newborn infant to detect gene product deficiencies and other abnormalities.

Nitrogenous base   This term refers to adenine (A), guanine (G), cytosine (C), and thymine (T). *See also* base; DNA.

Nuclear transfer   The basic procedure used in cloning organisms. It involves the removal of a cell's nucleus, which is then implanted into an animal's egg. Once it begins dividing, the new clone can be used either to harvest stem cells or to impregnate an animal. Nuclear transfer was the procedure Scottish scientists used to create the cloned sheep Dolly. (*See* cloning.)

Nucleic acid   A large molecule comprised of nucleotides. DNA and RNA are nucleic acids.

Nucleotide   One base of DNA or RNA (adenine, guanine, cytosine, thymine, or uracil), plus its associated phosphate and sugar molecules.

Nucleus   The cellular structure that houses most genetic material in the form of twenty-three pairs of chromosomes. (A tiny bit of genetic material is resident in another cellular structure, the mitochondria.)

Oncogene   A gene that normally acts as a growth regulator for cells. Many oncogenes are typically involved in cell growth. When overexpressed or mutated, they can cause cancer.

Oncology   The medical science devoted to the development, diagnosis, treatment, and prevention of tumors.

Organism   A living thing.

p53   *See* tumor-suppressor genes.

Parkinson's disease   A progressive neurological disease resulting in the death of brain cells associated with emotions and motor control.

Patent   Pertaining to DNA, a patent confers the right or title to a gene, gene variation, or an identifiable portion of genetic material to an individual or organization.

Pathogen   Any disease-causing organism.

Pathway   A system of proteins that work together. For example, a pathway may include one protein that sends a signal to a second protein, which sends a signal to a third protein, and so on, until a biological effect occurs.

Pedigree   A family tree diagram showing how a particular genetic trait or disease has been inherited through the generations.

Penetrance   A term that indicates the likelihood that a certain gene mutation will cause disease.

Peptide   A molecule made of two or more amino acids. Larger peptides are generally referred to as polypeptides or proteins.

Personalized medicine   The development of treatments and drug therapies specifically targeted to an individual's own genetic makeup.

Phage   A virus whose natural host is a bacterium.

Pharmacogenomics   The study of how an individual's genetic makeup will respond to specific drugs.

Phase I   A clinical trial that tests a new treatment in a small number of healthy people in order to determine its relative safety.

Phase II   A clinical trial that tests a new treatment in a small number of ill patients who have the disorder the drug is meant to treat. Phase II trials measure efficacy as well as safety.

Phase III   A clinical trial that tests a new treatment in a large number of ill patients who have the disorder the drug is meant to treat. Phase III trials measure efficacy as well as safety. This is the last phase of testing before a drug company applies for new drug approval (NDA) with the FDA.

Phenotype   The physical characteristics of an organism or the presence of a disease that may or may not be genetic. (*See also* genotype.)

Phenylketonuria (PKU)   A inherited error of metabolism that results in increased levels of the amino acid phenylalanine. Doctors frequently test newborns for PKU because untreated, it can lead to severe mental retardation.

Physical map   A diagram showing the locations of identifiable markers on DNA. Distance on the map is measured in base pairs. There are low-resolution and high-resolution maps. For the human genome, the lowest-resolution map shows all twenty-four chromosomes (the twenty-three plus the Y chromosome) and their colored banding patterns. The highest-resolution map shows the complete base sequence of every gene on every chromosome.

Plasmid   In a bacteria, the extra-chromosomal circular DNA that, in some cases, is capable of integrating into the genome of a host cell. In genetic engineering, plasmids are often used as vehicles to carry foreign "recombinant" DNA into a cell.

Pleiotropy   One gene that causes numerous physical traits—for example, multiple symptoms of a disease.

Pluripotency   A stem cell's potential to turn into more than one kind of mature cell, depending upon the environment surrounding it.

Polygenic disorder   A genetic disorder or disease that results from the combined and complex actions of a group of gene mutations. Examples of polygenic disorders include heart disease, diabetes, and some cancer. (*See* single-gene disorder.)

Polymerase chain reaction (PCR)   Invented by Kary B. Mullis, a method for amplifying a DNA base sequence so there is enough of a sample for scientists or technicians to test and otherwise examine. Sometimes called molecular photocopying, PCR is a key technology behind DNA fingerprinting and can create millions of copies from a single molecule.

Polymerase, DNA or RNA   The enzyme that catalyzes the creation of nucleic acids from preexisting DNA or RNA strands.

Polymorphism   A measurable difference in DNA sequences among individuals that may result in differences in health. Genetic variations occurring in more than one percent of a population are considered a useful polymorphism to analyze. (*See* linkage analysis; mutation.)

Polypeptide   A protein or part of a protein comprised of a chain of amino acids, each joined by a so-called peptide bond.

Population genetics   The study of how DNA sequences vary among a group of individuals.

Positional cloning   A technique used to identify genes based on their location on a chromosome.

Predictive gene tests   Tests that identify gene abnormalities that increase a person's likelihood of eventually suffering from certain diseases and disorders.

Prenatal diagnosis   The examination of fetal cells (taken from the amniotic fluid, placenta, or umbilical cord) for known biochemical and genetic mutations.

Probe   A specific sequence of single-stranded DNA, usually labeled with a radioactive atom or fluorescent dye, that is designed to bind to (and therefore single out) a particular strand of DNA that researchers are looking for.

Prokaryote   A cell or organism lacking a nucleus. Bacteria are examples of prokaryotes. (*See also* eukaryote.)

Proofreader genes   *See* DNA repair genes.

Prophylactic surgery   The removal of tissue that is in danger of becoming cancerous, before cancer has had the opportunity to develop. For instance, some women who test positive for BRCA1 and BRCA2 mutations predisposing them to breast cancer choose to remove their breasts, a procedure known as a prophylactic mastectomy.

Protein   A large, complex molecule, made of amino acids, that is essential to the structure, function, and regulation of every system in an organism's body. Examples of proteins are hormones, enzymes, and antibodies. The recipe for all a body's proteins is coded in the DNA.

Protein product   A specific protein molecule that is coded for and assembled by DNA in a particular gene.

Proteome   All the proteins expressed by the DNA in an individual's or population's genome.

Proteomics   The study of all the proteins a genome encodes.

Pseudogene   A sequence of DNA that looks like a gene but doesn't act like one; a pseudogene is probably a nonfunctional remnant of a once-functional gene that accumulated too many mutations over time.

RAC (Recombinant DNA Advisory Committee)   National Institutes of Health committee that advises the NIH director on whether to approve Public Health Service (PSA) supported gene therapy protocols.

Receptor   A protein or group of proteins in or on a cell that selectively binds to a substance (ligand). After it binds to a ligand, the receptor triggers a specific cellular response.

Recessive allele   A gene expressed only when its counterpart allele on the matching chromosome is also recessive, as opposed to dominant. Autosomal recessive disorders develop in persons who receive two copies of the mutant gene, one from each parent who is a carrier. This is also called a recessive gene. (*See also* dominant allele.)

Reciprocal translocation   The gene shuffling that results when a pair of chromosomes exchange exactly the same length and corresponding area of DNA.

Recombinant clone   A copy of a DNA sequence or organism that contains recombinant DNA. (*See* recombinant DNA molecules.)

Recombinant DNA molecules   The combination of DNA molecules of different origin.

Recombinant DNA technology   The procedures used to join together DNA segments of different origin—for instance, a sequence of human DNA in a strand of predominant bacterial DNA. This technology is also commonly referred to as genetic engineering and gene splicing.

Recombination   The process in which offspring end up with a combination of genes that is not exactly like either parent.

Regulatory region or sequence   A DNA base sequence that controls which genes on a given chromosome are expressed.

Repetitive DNA   Sequences of varying lengths that occur in multiple copies in the genome; repetitive DNA comprises much of the human genome.

Replication   The duplication of genetic material prior to cell division.

Reporter gene   *See* gene markers.

Reproductive cells   The egg (ovum) and sperm cells. Each mature reproductive cell carries a single set of twenty-three chromosomes. Other cells in the body, called somatic cells, include a double set of twenty-three chromosomes.

Resolution   The degree of detail on a physical map of DNA.

Restriction enzyme   A protein that recognizes specific, short base sequences and cuts DNA at exactly those sites. Bacteria contain over 400 such enzymes that recognize and cut more than 100 different DNA sequences. These are important tools in a genetic scientist's toolbox.

Restriction fragment length polymorphism (RFLP)   The variation between individuals in DNA fragment sizes because of specific restriction enzymes. RFLPs are often caused by mutations at a cutting site and are often useful to gene hunters as markers. (*See* polymorphism.)

Retinoblastoma   A type of eye cancer usually caused by the loss of a certain pair of tumor-suppressor genes. The inherited form typically appears in childhood, since one gene is missing from the time of birth.

Retroviral infection   The presence of RNA-containing viruses that use their RNA to create new DNA in a host cell. HIV is a retroviral infection.

Reverse transcriptase   An enzyme that retroviruses use to form a complementary DNA sequence (cDNA) from their RNA. The resulting DNA is then inserted into the chromosome of the host cell.

Ribonucleic acid (RNA)   A chemical found in the nucleus and cytoplasm of cells; it translates the protein-coding instructions of DNA into a code that the protein-building ribosomes of a cell can understand. The structure of RNA is similar to DNA—it also contains adenine (A), guanine (G), and cytosine (C), but instead of thymine (T), RNA contains uracil (U).

Ribosomes   The protein-building components of a cell; the sites of protein synthesis. (*See* ribonucleic acid.)

Risk communication   In the DNA sciences, the process of a genetic counselor interpreting and articulating genetic test results to a patient.

Sanger sequencing   Developed by Fred Sanger in the 1970s, this is a method of determining the order of bases in DNA. It is also sometimes called dideoxy sequencing. (*See also* sequencing; shotgun sequencing.)

Sarcoma   A type of cancer that starts in bone or muscle.

Screening   The effort of finding evidence of a particular disease in persons who have no symptoms of that disease.

Segregation   A normal biological process whereby the two pieces of a chromosome pair are separated during meiosis and randomly distributed to the germ cells.

Sequence assembly   The procedure of determining the order of multiple sequenced DNA fragments.

Sequence tagged site (STS)   A short (200 to 500 base pairs in length) DNA sequence that shows up once in the human genome and has a known location and sequence. Once detected by polymerase chain reaction (PCR), the STS is helpful as a marker and for reconciling different sequences from multiple labs. An expressed sequence tag (EST) is an STS derived from complementary DNA strands.

Sequencing   The process of determining the order of bases in a DNA molecule; the term can also refer to the determination of amino acids in a protein.

Sequencing technology   The instrumentation and procedures used to determine the order of bases in DNA.

Sex chromosomes   The X or Y chromosome in human beings; they determine the sex of an individual. Females have two X chromosomes. Males have an X and a Y chromosome. The sex chromosomes comprise the twenty-third chromosome pair in a karyotype. (*See* autosome.)

Sex-linked   Traits or diseases associated with the X or Y chromosome; generally seen in males.

Sexual reproduction   The production of offspring that occurs by combining the DNA from two individuals of the same species.

Shotgun sequencing   Method used by Celera Genomics to accelerate the sequencing of the human genome. It involves shredding up many cloned copies of the genome and randomly sequencing the pieces, with no advanced knowledge of where the segment originally belongs on the genome. A high-powered computer then arranges the sequence of the bases in the most likely order. This is opposed to so-called directed sequencing, in which pieces of DNA from known chromosomal locations are sequenced. (*See also* library.)

Sickle cell anemia   An inherited, sometimes lethal disease in which a defect in hemoglobin changes the shape (sickling) and loss of red blood cells, resulting in damage to organs throughout the body.

Single-gene disorder   A hereditary disorder caused by a mutant allele of a single gene. Some examples include sickle cell disease and retinoblastoma.

Single nucleotide polymorphism (SNP)   The DNA sequence variation that occurs when a single nucleotide (A, T, C, or G) in the genome sequence is altered.

Small molecule drug   One or more active chemical compounds, usually formulated as a pill, which interact with a specific biological target to provide a curative effect.

Somatic cell   Any cell in the body except the sex cells and their precursors.

Somatic cell gene therapy   Experimental technique of incorporating new genetic material into cells in order to treat disorders and abnormalities. Because no new material is going to the sex cells, the genetic changes cannot be passed to the patient's offpring. Somatic cell gene therapy is the only kind of gene therapy approved by the FDA. (*See* gene therapy.)

Somatic cell genetic mutation   *See* acquired mutations.

Somatic mutations   *See* acquired mutations.

Sporadic cancer   A cancer that is not inherited from the parents, but is a result of a random genetic mutation. It is caused by DNA changes in one cell that grows and divides out of control, later spreading throughout the body. (*See* hereditary cancer.)

Stem cell   Undifferentiated, early cells in the bone marrow and in embryos that have the ability both to multiply and to grow into different kinds of mature, specialized cells.

Structural genomics   The field of determining the three-dimensional structures of proteins through experimental techniques and computer simulation.

Substitution In the DNA sciences, a type of mutation resulting from the replacement of one base in a DNA sequence with another one. May also refer to the replacement of an amino acid in a protein by another amino acid. (*See* mutation.)

Suppressor gene A gene that, if expressed, can suppress the action of another gene.

Syndrome A pattern of symptoms or abnormalities that indicates a certain trait or disorder.

Tandem repeat sequences The same base sequence occurring in multiple spots on a chromosome; useful as markers in physical mapping. (*See* physical map.)

Targeted mutagenesis A laboratory-induced mutation in a genetic sequence at a specific site on a chromosome; used by researchers to discover a certain region's function. (*See* mutation; polymorphism.)

Tay-Sachs disease An inherited disease among infants causing severe mental retardation and early death. Caused by a recessive gene mutation, it is especially common in those of Eastern European Jewish descent.

Technology transfer Taking scientific findings from research labs into companies—or from biotech companies to pharmaceutical companies—where they can be commercialized for purposes of making a profit.

Telomerase The enzyme responsible for directing the replication of telomeres. (*See* telomere.)

Telomere The special structure at the end of every chromosome. It is involved in the continued replication of DNA. (*See* DNA replication.)

Teratogenic Radiation, toxins, or other chemicals that cause embryos to develop abnormally. (*See* mutagen.)

Thymine (T) One of the four bases in DNA, always pairing with adenine (A). *See also* base pairs; nucleotide.

Transcription The copying of information from DNA into new strands of messenger RNA (mRNA). The mRNA then carries this information to the ribosomes, where it serves as a recipe for putting together the amino acids needed to build a protein.

Transfer RNA (tRNA) A type of RNA involved in protein synthesis, specifically with bonding to amino acids and transferring them to the ribosomes, where proteins are put together based on the code carried in the messenger RNA. (*See* mRNA.)

Transformation   How an individual cell's genome is altered by adding DNA from a cell foreign (i.e., from another organism) to it.

Transgenic   The artificially created organism that results when scientists incorporate foreign DNA into an organism's germ line. (*See* germ line.)

Translation   The process of taking instructions from messenger RNA (mRNA), base by base, into chains of linked amino acids that eventually fold into proteins. This process takes place outside of the nucleus, on cellular structures called ribosomes.

Translocation   A large segment of one chromosome breaking off and attaching to another chromosome. (*See* mutation.)

Transposable element   DNA sequences that are potentially able to jump from one chromosomal site to another.

Trisomy   The possession of three copies (instead of two) of a particular chromosome. For example, trisomy 13 is a severe birth defect caused by the inheritance of three copies of human chromosome 13. (*See* chromosome.)

Tumor-suppressor genes   Specific genes on the genome that normally monitor and restrain cell growth. When mutated or missing, they allow cells to grow uncontrollably, leading to tumors.

Uracil (U)   One of the four bases in RNA, replaces thymine (T) and always bonds with adenine (A). For example, the sequence GAT in DNA will always translate to GAU in RNA. (*See* base; nucleotide.)

Vector   A DNA molecule—often originating from a virus or cell of a higher organism—that is used as a vehicle to introduce foreign DNA into cells. An important tool in gene therapy, a vector is most frequently a recombinant DNA molecule produced in the lab that includes DNA from two or more sources.

Virus   A simple organism essentially consisting of DNA or RNA covered with protein. Viruses cannot live on their own; they survive by transmitting their genetic material into the hosts they infect, thereby reproducing themselves. (*See* vector.)

Wild type   The variety of an organism that occurs most commonly in nature.

Working draft DNA sequence   The draft of the DNA sequence announced by Celera Genomics and the Human Genome Project in June 2000. The final and complete version of the DNA sequence was announced in April 2003.

X chromosome   One of the two sex chromosomes. (*See also* Y chromosome.)

Y chromosome   One of the two sex chromosomes. (*See also* X chromosome.)

Yeast artificial chromosome (YAC)   Created from the DNA of yeast, this is a vector commonly used to copy (clone) large fragments of DNA. (*See* vector.)

Zinc-finger protein   A notable feature of some proteins containing a zinc atom.